Quantum Computing for Engineers

Osama M. Raisuddin · Suvranu De

Quantum Computing for Engineers

Springer

Osama M. Raisuddin
Future of Computing Institute
Rensselaer Polytechnic Institute
Troy, NY, USA

Suvranu De
Florida A&M University-Florida State
University College of Engineering
Tallahassee, FL, USA

ISBN 978-3-032-03324-6 ISBN 978-3-032-03325-3 (eBook)
https://doi.org/10.1007/978-3-032-03325-3

This Springer imprint is published by the registered company Springer Nature Switzerland AG
The registered company address is: Gewerbestrasse 11, 6330 Cham, Switzerland

Preface

Quantum computing is a rapidly emerging field. As researchers entering the field from outside traditional domains, such as theoretical physics or computer science, we found the landscape both rich and fragmented—dense with promise but difficult to navigate without a cohesive, application-oriented framework. This book was born from our own attempts to cross that threshold from classical computational methods into quantum computing—not as physicists, but as computational engineers—and to build a practical bridge for others who wish to do the same.

While several introductory texts exist, they often assume prior exposure to quantum theory or computational complexity. Our aim is different: to build a bridge from classical engineering and scientific computing to quantum algorithms, without sacrificing technical depth. This book adopts an applied, algorithmic perspective, integrating mathematical foundations, computational models, and real-world applications.

The book is organized into seven parts, each divided into focused chapters. These chapters are intended to be modular and self-contained, allowing instructors or readers to adapt them to different learning paths or interests:

- **Part I: Mathematical and Computational Preliminaries**
  Foundational material in linear algebra, probability, and numerical methods, tailored to the quantum context.
- **Part II: A Brief Introduction to Quantum Mechanics**
  Core quantum phenomena presented through key experiments, emphasizing physical intuition.
- **Part III: Elements of Quantum Computing**
  The computational framework of quantum computing—qubits, gates, circuits, and measurement.
- **Part IV: Programming Quantum Computers**
  Practical aspects of working with quantum systems, including software stacks, noise, and available libraries.
- **Part V: Algorithmic Primitives, Subroutines, and Frameworks**
  Reusable algorithmic components that form the building blocks of more advanced methods.

- **Part VI: Quantum Algorithms**
  Complete quantum algorithms for solving problems in linear algebra, differential equations, and optimization.
- **Part VII: Applications, Future Directions, and Open Problems**
  Case studies in engineering and finance, along with discussions on open challenges and research frontiers.

Each chapter develops a coherent topic, moving from concept to method to application, often supported with example code or pseudocode. Conceptual explanations are supplemented with figures, algorithmic breakdowns, and Python-based implementations using Qiskit. Where relevant, mathematical derivations clarify the underlying logic, and code examples reinforce the connection between theory and practice. Our goal is to equip readers with both the theoretical foundation and practical tools necessary to engage with quantum computing in an engineering context, whether for research, development, or curriculum design.

By the end of this book, readers will have a solid understanding of the principles of quantum computation, be able to model and implement core quantum algorithms, and critically assess where quantum methods can offer computational advantages. They will also gain familiarity with quantum programming environments and be prepared to pursue further work in both applied and theoretical directions.

The book assumes familiarity with undergraduate-level linear algebra, probability theory, algorithmic reasoning, and computational problem-solving. Prior exposure to quantum mechanics is not required; the essential physical principles are introduced in Part II. Familiarity with Python programming is recommended for engaging with the hands-on components.

All code examples in this book use Python and the Qiskit software stack. Python was selected for its widespread use in scientific and engineering computing, and Qiskit offers a well-supported platform for constructing and simulating quantum circuits. While Qiskit adopts little-endian indexing for qubits, we retain big-endian notation in equations and figures for mathematical clarity. These conventions are made explicit where relevant.

Supplementary code and figures are available at https://github.com/osamarais/QuantumComputingForEngineers. These resources are designed to support both independent learners and instructors using this book in the classroom.

Suggested Paths for Different Readers

This book is intended to be accessible to a broad engineering and scientific audience with a background in linear algebra and basic programming. Readers with different goals may choose to navigate it differently:

- **For engineering students and newcomers to quantum computing**, Parts I–III provide the necessary foundation, and Parts IV–V introduce core programming and algorithmic tools.
- **For researchers and advanced practitioners**, Parts V–VII offer in-depth treatments of quantum algorithmic frameworks and applications, with references to the underlying theory where needed.
- **For instructors**, each chapter can serve as a standalone module in an advanced undergraduate- or graduate-level course. The modular structure allows for flexible integration into existing curricula in computing, applied mathematics, or engineering.

We hope this book serves as both a foundation and a launchpad for those beginning their journey into quantum computing, and for those aiming to apply it meaningfully in engineering and scientific domains.

Troy, NY, USA — Osama M. Raisuddin
Tallahassee, FL, USA — Suvranu De

Contents

Part V Algorithmic Primitives, Subroutines, and Frameworks

Part VI Quantum Algorithms

Part I
Mathematical and Computational Preliminaries

This part provides a focused overview of the mathematical foundations required to engage with quantum computing in the context of this book. The Dirac notation is adopted from the outset, and concepts are introduced in chapters with an emphasis on their relevance to quantum algorithms, rather than as exhaustive treatments.

The first few chapters provide an overview of some key mathematical concepts required for the remainder of this book. These chapters are meant as a refresher, not a comprehensive course. Readers seeking a deeper dive into mathematics may wish to consult standard texts in those areas. Our focus is on presenting just enough background to support the development of later topics. We also provide context and motivation for each concept, highlighting connections to quantum computing or quantum physics. Readers are encouraged not to dwell too deeply on these connections at this stage; many will be revisited and expanded upon in later parts. This part builds on the following chapters.

Chapter 1, "Linear Algebra and Probability", presents the essential mathematical tools—linear algebra and probability—that form the backbone of quantum computation.

Chapter 2, "Polynomial Approximations", introduces techniques for approximating functions with polynomials, both at a point and over an interval, which are foundational for understanding several advanced quantum algorithms.

Chapter 3, "Theory of Computing", introduces fundamental concepts from classical computation, including Turing machines and universal circuit families, along with examples of computable and incomputable problems.

Chapter 4, "An Overview of Practical Classical Computing", connects abstract models to hardware, covering transistors, logic gates, and the translation from high-level code to machine instructions.

Chapter 5, "Information and Complexity Theory", introduces classical complexity classes (P, NP, and BPP) and the quantum class BQP, motivating quantum speedups while grounding expectations. A concluding thought experiment links information theory to thermodynamics.

1 Linear Algebra and Probability

Linear algebra and probability are foundational requirements for quantum computing and are sufficient for understanding and analyzing basic concepts and algorithms in quantum computing. This chapter is a brief overview and refresher of linear algebra and probability for the remainder of the book.

We first introduce column and row vectors in the notation familiar to engineers and then transition to the Dirac or "bra-ket" notation used in quantum computing, which is the standard notation used in quantum computing literature.

Subsequently, important matrix classes such as Hermitian, unitary, and orthogonal projection matrices are introduced with explanations of how they represent operations on a quantum state. These concepts are then connected to the expectation values of a quantum state and the properties of valid expectation value operators.

We then introduce the concept of measurements, which are mathematically expressed as projection operations in the simplest case and explain how measurements correspond to sampling outcomes from a probability distribution determined by the probability amplitudes of a quantum state.

We then transition to numerical problems involving matrices. We provide an overview of important matrix decompositions, including the eigenvalue transformation to diagonalize a matrix and the singular value decomposition, along with the commonly encountered eigenvalue problem. The condition number of matrices is shown to be an important property of matrices, which generally has implications for the cost and precision of a solution.

The linear system of equations problem is presented with a discussion of two major classes of algorithms to solve it: direct solvers and iterative solvers.

We end the chapter by introducing the Krylov subspace and Kronecker product of matrices with some relevant properties.

O. M. Raisuddin and S. De, *Quantum Computing for Engineers*,
https://doi.org/10.1007/978-3-032-03325-3_1

Vectors, Bras, Kets, and Dirac Notation

A vector $v \in \mathbb{C}^N$ is an element of an N-dimensional linear space of complex vectors, $\mathbb{C}^N$, with entries $v_i \in \mathbb{C}$, and can be written as $v = \sum_{i=0}^{N-1} v_i e_i$ where e_i is the $i\,th$ standard basis vector.

In quantum mechanics, vectors are typically expressed using Dirac notation, also known as *bra-ket* notation. A "ket," written as $|\psi\rangle$, represents the state of a quantum system. In finite-dimensional systems like those in gate-based quantum computing, a ket is simply a complex column vector in $\mathbb{C}^N$.

Quantum states are normalized, meaning that their squared magnitude is 1. Kets can be written in any orthonormal basis, and the label $|\psi\rangle$ denotes a specific state in that basis. The most common basis in quantum computing is the computational basis, consisting of states $|0\rangle, |1\rangle, \ldots, |N-1\rangle$, which correspond to the standard basis vectors in $\mathbb{C}^N$, e.g.,

$$|0\rangle = \begin{pmatrix} 1 \\ 0 \end{pmatrix}, |1\rangle = \begin{pmatrix} 0 \\ 1 \end{pmatrix} \in \mathbb{C}^2$$

This basis is sometimes called the Z-basis, as it is the set of eigenstates of the Pauli-Z operator.

Qubits are the quantum computational analog of classical bits, i.e., 0 or 1. Another commonly used label is an integer $i \in \mathbb{W}$ which represents a quantum state corresponding to the $i\,th$ standard basis vector

$$|i\rangle = e_i$$

Note that this notation does not conflict with the notation introduced earlier for computational bits when $i = 1$ or 0. An arbitrary vector v may be represented in Dirac notation (after normalization) as

$$|v\rangle = \frac{1}{\sqrt{v^\dagger v}} \sum_{i=0}^{N-1} v_i |i\rangle$$

where the "dagger" symbol $\dagger$ denotes the conjugate transpose operation.

A "bra" $\langle\psi|$ is the adjoint of a ket $|\psi\rangle$ and can be represented as the row vector $\langle\psi| = (|\psi\rangle)^\dagger$. For brevity, we often refer to an arbitrary bra or ket as $\langle\psi|$ and $|\psi\rangle$, respectively, with elements $\psi_i \in \mathbb{C}$ where normalization is implied:

$$|\psi\rangle = \sum_i \psi_i |i\rangle, \ \sum_i |\psi_i|^2 = 1$$

Bras and kets must be normalized because they represent "probability amplitudes" and, by extension, a probability distribution (and probabilities of all event outcomes must sum to 1). This normalization is often referred to as the Born rule.

Probability amplitudes are different from probabilities. Probability amplitudes are complex numbers $\psi_i \in \mathbb{C}$ in the unit ball, i.e., $|\psi_i| \leq 1$, whereas probabilities are real numbers $p_i \in \mathbb{R}$ s.t. $0 \leq p_i \leq 1$. The probability corresponding to a probability amplitude may be computed as $\psi_i^* \psi_i$. Unlike probabilities, probability amplitudes allow quantum states to interact constructively or destructively.

Quantum states such as $|v\rangle$ and $|w\rangle$ belong to a Hilbert space, a complex vector space equipped with an inner product. The inner product between two kets $|v\rangle$ and $|w\rangle$ is written as $\langle v \mid w\rangle$ and computed as

$$\langle v \mid w\rangle = \sum_{i=0}^{N-1} v_i^* w_i$$

This inner product corresponds to the overlap between the states. Physically, the squared magnitude $|\langle v \mid w\rangle|^2$ gives the probability of obtaining an outcome $|v\rangle$ when measuring a system in a state $|w\rangle$.

Note: this requires definition of measurements, and the basis used for measurement, which is explained in detail in Chap. 10: Quantum Measurements and Circuits.

We note that if the bra and ket represent quantum objects in a continuous Hilbert space, they can be represented as functions over a support ω

$$|v\rangle = v(\omega)$$

$$\langle w| = w^*(\omega)$$

for which a bra-ket represents an integral:

$$\langle v \mid w\rangle = \int v(\omega)^* w(\omega) d\omega$$

and the bra and ket are normalized, i.e.,

$$\langle v \mid v\rangle = \int v(\omega)^* v(\omega) d\omega = 1$$

$$\langle w \mid w\rangle = \int w(\omega)^* w(\omega) d\omega = 1$$

To further explain the interpretation of quantum states as probability distributions, we first need to introduce matrices and matrix operations on quantum states.

Linear operators acting on a Hilbert space may be expressed in terms of an outer product. Let $|v\rangle$ and $\langle w|$ belong to Hilbert spaces V and W, respectively.

Their outer product, written as $|w\rangle\langle v|$, represents a linear operator from V to W, defined as

$$(|w\rangle\langle v|)\left|v'\right\rangle = \left(\left\langle v|v'\right\rangle\right)|w\rangle \ \forall \ v' \in \mathrm{V}$$

If $|v\rangle = \sum_i v_i|i\rangle$ in the standard basis, where $v_i = \langle i \mid v\rangle$, then

$$|v\rangle = \sum_i |i\rangle\langle i \mid v\rangle = \sum_i (|i\rangle\langle i|)|v\rangle = \left(\sum_i |i\rangle\langle i|\right)|v\rangle$$

indicating that the identity matrix $I = \sum_i |i\rangle\langle i|$ as the above relationship holds for all $|v\rangle$.

Similarly, if $|w\rangle = \sum_j w_j|j\rangle$, then $|w\rangle\langle v| = \sum_i\sum_j v_i w_j|i\rangle\langle j|$ defines the matrix (linear operator) A with elements $A_{ij} = v_i w_j$.

As an example, consider a general qubit state

$$|\psi\rangle = \alpha|0\rangle + \beta|1\rangle = \begin{pmatrix} \alpha \\ \beta \end{pmatrix}$$

and a linear operator constructed from outer products of the standard basis states:

$$\begin{aligned} A &= a|0\rangle\langle 0| + b|0\rangle\langle 1| + c|1\rangle\langle 0| + d|1\rangle\langle 1| \\ &= a\begin{pmatrix} 1 & 0 \\ 0 & 0 \end{pmatrix} + b\begin{pmatrix} 0 & 1 \\ 0 & 0 \end{pmatrix} + c\begin{pmatrix} 0 & 0 \\ 1 & 0 \end{pmatrix} + d\begin{pmatrix} 0 & 0 \\ 0 & 1 \end{pmatrix} \\ &= \begin{pmatrix} a & b \\ c & d \end{pmatrix} \end{aligned}$$

Applying A to $|\psi\rangle$ yields

$$A|\psi\rangle = \begin{pmatrix} a & b \\ c & d \end{pmatrix}\begin{pmatrix} \alpha \\ \beta \end{pmatrix} = \begin{pmatrix} \alpha a + \beta b \\ \alpha c + \beta d \end{pmatrix} = (\alpha a + \beta b)|0\rangle + (\alpha c + \beta d)|1\rangle$$

Matrices

A matrix A is an operator that can be applied to a vector to perform the map

$$w = Av$$

where $w, v \in \mathbb{C}^N, A \in \mathbb{C}^{N\times N}$.

When operating on a quantum state $|\psi\rangle$, it is necessary to maintain the normalization of the quantum state according to the Born rule. Unitary matrices ensure that the Born rule is not violated. By definition, a unitary matrix U satisfies

$$UU^{\dagger} = I$$

Unitary matrices $U \in \mathbb{C}^{N\times N}$ belong to the mathematical group $U(N)$, referred to as a unitary group of degree N. Unitary matrices $U \in \mathbb{C}^{N\times N}$ such that $\det(U) = 1$ (the determinant) belong to the mathematical group $SU(N)$, referred to as the special unitary group of degree N. Unitary matrices have several special properties:

- Unitary matrices are 2-norm preserving, i.e., $\|U|\psi\rangle\|_2 = 1\ \forall\ U \in U(N), |\psi\rangle \in \mathbb{C}^N$.
- Unitary matrices are normal, i.e., $UU^{\dagger} = U^{\dagger}U = I$.
- The eigenvalues $\lambda_i(U)$ of a unitary matrix satisfy $|\lambda_i(U)| = 1\ \forall\ i$, and therefore $|\det(U)| = 1$.
- A unitary matrix is diagonalizable as $U = VDV^{\dagger}$ where V and D are also unitary.
- The product of two unitary matrices is also unitary.
- Any unitary matrix can be written as a (non-unique) exponentiated Hermitian matrix $U = e^{iH}$.

Unitary matrices are important since they mathematically represent the manipulation of a quantum state as a matrix–vector multiplication operation (without performing any sampling from the probability distribution or a measurement of the physical entity it represents). A particularly important property is that they are 2-norm preserving. If a ket represents a probability distribution, it must be normalized, and any manipulation of the ket must ensure $\||\psi\rangle\|_2 = 1$, which is automatically satisfied by unitary matrices.

Hermitian matrices are another important class of matrices for quantum computing. By definition, a Hermitian matrix $H \in \mathbb{C}^{N\times N}$ satisfies

$$H = H^{\dagger}$$

Hermitian matrices also possess some important properties:

- The eigenvalues $\lambda_i(H)$ of a Hermitian matrix satisfy $\lambda_i(H) \in \mathbb{R}\ \forall\ i$
- The exponentiation e^{iH} of a Hermitian matrix is a (unique) unitary matrix
- A Hermitian matrix is unitarily diagonalizable
- The quadratic form of a Hermitian matrix $v^T Hv \in \mathbb{R}\ \forall\ v \in \mathbb{C}^N$ is real $\big(\langle\psi|H|\psi\rangle \in \mathbb{R}\ \forall\ |\psi\rangle \in \mathbb{C}^N\big)$.

Hermitian matrices play several important roles in quantum computing and quantum mechanics. They define "observables" of a quantum object. An observable may be thought of as a measurable property of a physical object. In the context of quantum objects, an example would be whether an electron is observed to have a "spin up" or "spin down" configuration. Although quantum objects are mathematically represented in complex space, in the classical world we interact with them and measure them as real numbered quantities; therefore, it does not make sense for a single "observation" or "measurement" of any physical object to be a complex number; the observation or measurement is always real. This requirement is automatically satisfied by the fact that the quadratic form of Hermitian matrices is real for any $|\psi\rangle$. Mathematically, observables are defined as operators on bras and kets. The concept of observables and measurements will be revisited in Chap. 6, A Gentle Introduction to Quantum Mechanics, when introducing the postulates of quantum mechanics.

Hermitian matrices are therefore suitable for defining observables of a quantum object. The expectation value of an observable with an operator H corresponding to a quantum state $|\psi\rangle$ is mathematically defined as

$$\langle H \rangle_\psi = \langle \psi | H | \psi \rangle$$

This can be understood as follows: If a quantum mechanical system is known to be in some state $|\psi\rangle$, a series of measurements of an observable property (which corresponds mathematically to the observable operator H) will have a mean/expectation value $\langle H \rangle_\psi$.

Matrices can also be defined as positive-definite or positive-semidefinite. This is important for mathematically defining the energy of a physical object as a Hamiltonian $\mathcal{H}$. A Hamiltonian of a physical system is a mathematical operator that returns the total energy of the system. As an example, the total energy E (an observable) of a quantum state $|\psi\rangle$ is computed using the Hamiltonian $\mathcal{H}$ (operator for the observable E) as

$$E = \langle \psi | \mathcal{H} | \psi \rangle$$

Rotation and Reflection Matrices

Unitary matrices in $U(N)$ can represent rotations and reflections, while the subset $SU(N)$ only represents rotations. Rotation and reflection matrices can be distinguished by their determinants: $\det(U) = 1$ for rotation matrices while $\det(U) = -1$ for reflection matrices.

Pauli Matrices and Pauli Basis

Like vectors being expressed in terms of a set of basis vectors, matrices may also be represented in a matrix basis. A particularly important and useful basis for matrices is the Pauli basis formed by the Pauli matrices:

$$\sigma_x = \begin{pmatrix} 0 & 1 \\ 1 & 0 \end{pmatrix}, \ \sigma_y = \begin{pmatrix} 0 & -i \\ i & 0 \end{pmatrix}, \ \sigma_z = \begin{pmatrix} 1 & 0 \\ 0 & -1 \end{pmatrix}$$

Later in Chap. 9, Qubits, Quantum Registers, and Quantum Gates, we redefine these Pauli matrices as single-qubit gates.

Together with the identity matrix (I), the Pauli matrices form a complete basis for 2×2 matrices. Any such matrix can be written as

$$A_{2\times2} = \alpha_1 I + \alpha_2\sigma_x + \alpha_3\sigma_y + \alpha_4\sigma_z$$

where $\alpha_i \in \mathbb{C}$. This result can be extended to any general matrix $A \in \mathbb{C}^{2n\times2n}$ as

$$A = \frac{1}{2^n} \sum_{i_1,i_2,\ldots,i_n} \alpha_{i_1,i_2,\ldots,i_n} \sigma_{i_1} \otimes \sigma_{i_1} \otimes \ldots \otimes \sigma_{i_n}$$

where $\sigma_{i_j} \in \{I, \sigma_x, \sigma_y, \sigma_z\}$, $\alpha_{i_1,i_2,\ldots,i_n} \in \mathbb{C}$, and $\otimes$ denotes the Kronecker product. Furthermore, we note that one may always express an arbitrary matrix as a Hermitian matrix using the Hermitian dilation of A:

$$H = \begin{pmatrix} 0 & A \\ A^\dagger & 0 \end{pmatrix}$$

Chapter 13, Access Models and Data Representation, provides an example code to decompose matrices in the Pauli basis.

Vector and Matrix Norms

Vectors and matrix norms are important metrics used in quantum computing. A p-norm of a vector for $p \in \mathbb{Z}^+$ is defined as

$$\|v_p\| = \left(\sum_i (v_i)^p\right)^{1/p}$$

Similarly, a norm may be defined for a matrix as an induced p-norm:

$$\|A\|_p = \sup_{x\neq0}\left\{\frac{\|Ax\|_p}{\|x\|_p}\right\}$$

Of particular interest is the 2-norm, i.e., $p = 2$. In this book, whenever the norm is unspecified, it may be assumed to be the 2-norm.

Condition Number

The condition number of a matrix A is defined as

$$\kappa(A) = \sup_{v,w} \left\{ \frac{\|A^{-1}v\|}{\|v\|} \frac{\|w\|}{\|A^{-1}w\|} \right\} = \|A\| \|A^{-1}\| \geq 1$$

The condition number of a matrix frequently arises in the analysis of quantum algorithms involving matrices, such as linear system solvers or ordinary differential equation solvers. A large condition number will often manifest itself as difficulties in solving a problem, either through a loss of digits of precision or the overall cost of a solver.

We note the following important properties:

- An equivalent definition of $\kappa(A)$ is the ratio of the singular values $\sigma_{max}/\sigma_{min}$
- For normal matrices (e.g., Hermitian and unitary matrices), $\kappa(A)$ is equivalently defined as a ratio of the eigenvalues $\|\lambda_{max}\|/\|\lambda_{min}\|$
- $\kappa(A) = 1$ for unitary matrices.

For diagonalizable matrices $A = V\Lambda V^\dagger$, ill-conditioning may arise from the ratio of the eigenvalues and/or the eigenvectors being approximately linearly dependent.

Projectors

An orthogonal projector $P \in \mathbb{C}^{N\times N}$ is a matrix such that

$$P^2 = P$$

and

$$P^\dagger = P$$

We refer to orthogonal projectors simply as projectors (as opposed to non-orthogonal or oblique projectors).

Note: Orthogonal projectors are not orthogonal matrices.

A projection matrix maps a vector $v \in \mathbb{C}^N$ to a subspace $\mathbb{C}^M$ s.t. $M \leq N$ and therefore rank$(P) \leq N$, as shown in Fig. 1.1. Projectors are important for defining measurement operators in quantum computing. The eigenvalues of orthogonal projectors are $\lambda(P) \in \{0, 1\}$.

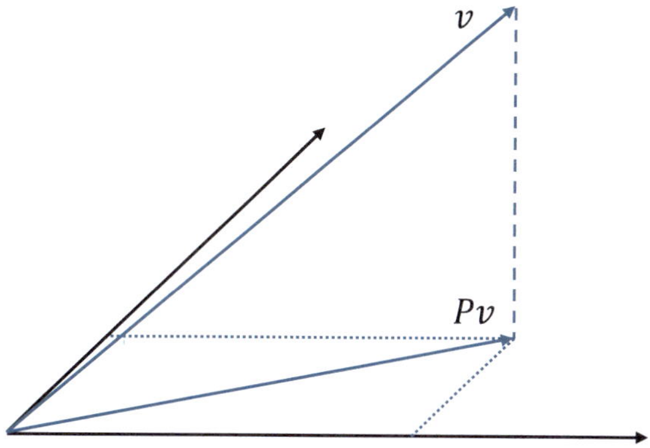

Fig. 1.1 An orthogonal projector $P \in \mathbb{R}^{3\times3}$ projecting a vector $v \in \mathbb{R}^3$ onto Pv

Eigenvalue Problems

Given a matrix A, an eigenvalue problem in linear algebra can be defined as.

Find v, λ such that

$$Av = \lambda v$$

where v is an eigenvector and λ is an eigenvalue.

In matrix form this can be written as

$$AV = \Lambda V$$

where Λ is a diagonal matrix with the eigenvalues lying on the diagonal, and V is a matrix whose columns form the eigenvectors of A.

A matrix may have a full set of eigenvectors, i.e., a unique eigenvector for each eigenvalue, in which case it can be diagonalized uniquely as

$$A = V\Lambda V^{\dagger}$$

If a matrix is Hermitian, it has orthogonal eigenvectors with real eigenvalues. However, not every normal matrix is Hermitian (e.g., unitary matrices are normal but can have complex eigenvalues such that $|\lambda_i| = 1 \ \forall \ i$). Furthermore, not every non-singular matrix is diagonalizable; e.g., if a matrix has repeated eigenvectors, then it can be diagonalized using the Jordan normal form (e.g., a triangular matrix of 1's).

Note that non-singular Hermitian matrices commute, i.e., $[A, B] = AB - BA = 0$, *iff* they share the same eigenvectors.

For normal matrices $A_\perp$, one may define the condition number as

$$\kappa(A_\perp) = \frac{|\lambda_{max}|}{|\lambda_{min}|}$$

This formula does not hold for non-normal matrices since the eigenvectors themselves may be ill-conditioned, i.e., they are not mutually orthogonal and $\kappa(V) > 1$.

It is important to note that, in general, for matrices with $N > 4$ a general formula or deterministic procedure to compute eigenvalues does not exist. This is because eigenvalues are the roots of the characteristic polynomial

$$f(\lambda) = \det(A - \lambda I)$$

where $\det(\cdot)$ denotes the determinant and $f(\cdot)$ is a polynomial, and a closed-form algebraic formula to compute the roots of a polynomial of degree > 4 has been shown not to exist according to the Abel–Ruffini theorem. Therefore, the computation of eigenvalues and eigenvectors in general must be iterative.

We now briefly introduce the generalized eigenvalue problem. Given two symmetric matrices A and S (sometimes referred to as a linear matrix pencil (A, S)), a generalized eigenvalue problem can be defined as follows.

Find v, λ *such that*

$$Av = \lambda S v$$

where v is a generalized eigenvector and λ is a generalized eigenvalue.

The standard eigenvalue problem is a special case of the generalized eigenvalue problem with $S = I$. S is often referred to as an overlap matrix.

Eigenvalue problems are of particular interest in engineering since they can be used to study the stability of dynamic systems and instabilities (e.g., buckling) in static systems, solve ordinary differential equations, and tackle many other problems of practical interest.

Singular Value Decomposition

The singular value decomposition (SVD) of a matrix is defined as

$$A = U \Sigma V^{\dagger}$$

where U and V are matrices, whose columns are the left and right singular vectors, respectively, and Σ is a diagonal matrix with the singular values $\sigma \in [0, \infty) \cong \mathbb{R}^{+}$ lying on the diagonal, ordered from the largest to the smallest. The SVD is unique up to the signs of the singular vectors.

The SVD is particularly powerful since it applies to every matrix, whether square or rectangular, singular or non-singular. The norm of any matrix and its inverse are

$$\|A\| = \sigma_{max}, \ \left\|A^{-1}\right\| = \sigma_{min}$$

Therefore, the condition number of a matrix is directly available from an SVD as

$$\kappa = \frac{\sigma_{max}}{\sigma_{min}}$$

The "most important subspace" of a matrix (and its inverse) containing the "bulk of the information," in the 2-norm, corresponds to the largest (and the smallest) singular values. Furthermore, inverting a matrix using its SVD is particularly straightforward, requiring two dense matrix multiplications and one diagonal matrix multiplication at most.

The singular value decomposition is crucial for advanced quantum algorithmic frameworks, such as the quantum singular value transformation and quantum signal processing.

Linear Systems of Equations

A linear system problem, or a system of linear equations, is defined as.

Given $A \in \mathbb{C}^{N \times N}$, $b \in \mathbb{C}^N$, *find* $x \in \mathbb{C}^N$ *s.t.*

$$Ax = b$$

The linear system problem has a solution, or is well-posed, *iff* $\det(A) \neq 0$ or equivalently:

All singular values $\sigma > 0$.

All eigenvalues $\lambda \neq 0$.

A^{-1} exists.

The linear system problem arises frequently in science and engineering. Robust and efficient solutions of linear systems are of paramount importance in scientific computing. There exist two broad classes of algorithms for solving systems: *direct solvers* and *iterative solvers.*

Direct solvers for linear systems have a runtime that is known *a priori,* but have the disadvantage of scaling worse than iterative solvers in computational cost. Furthermore, direct solvers only provide a solution at the end of the computation, whereas iterative solvers can typically provide a useful partially converged solution before fully converging.

Classical direct solvers, e.g., LU decomposition, require $O(N^3)$ operations in general to solve a dense linear system with N unknowns. Direct solvers with better asymptotic scaling have been discovered but are not practically implementable or useful due to large prefactors.

Iterative solvers scale more favorably. Given an initial guess with error $\|b - Ax_0\| = \epsilon_0$, the conjugate gradient iterative solver scales as $O(\sqrt{\kappa}N \log N/\tilde{\epsilon})$ for positive-definite A and $O(\sqrt{\kappa}N \log N/\tilde{\epsilon})$ for indefinite A where $\tilde{\epsilon} = \epsilon_0/\epsilon$ is the relative decrease in solution error. More advanced iterative solvers, e.g., multigrid methods, require as little as $O(N \log \tilde{\epsilon})$ floating-point operations (FLOPs).

Table 1.1 Some common linear stationary iterative methods

Method	Iterative formula	
Richardson	$x_{n+1} = x_n + \omega(b - Ax_n)$	ω is a scalar parameter chosen for convergence
Jacobi	$x_{n+1} = D^{-1}(b - (L + U)x_n)$	D, L, U are the diagonal, upper triangular, and lower triangular parts of A
Gauss–Seidel	$Lx_{n+1} = b - Ux_n$	

The crux of effective and efficient iterative solvers is exploiting the structure of the problem underlying the linear system.

Quantum algorithms for linear systems have the potential to achieve an exponential speedup in N. However, they scale linearly with κ at best. Improving the linear dependence of κ for specific problems of practical interest, e.g., sparse positive-definite matrices, is an active area of research. Quantum linear system algorithms are discussed in detail in Chap. 30, Quantum Linear System Algorithms: Direct Methods, and Chap. 31, Quantum Linear System Algorithms: Iterative Methods.

Linear Iterative Methods: Stationary Point Iterations

A simple and straightforward way of solving linear systems iteratively is using a linear iteration whose stationary point is x.

These iterations take the general form:

$$x_{n+1} = f(x_n)$$

where $f(x)$ is an affine transformation. Richardson, Jacobi, and Gauss–Seidel methods are well-known and elementary techniques for solving linear systems iteratively and are summarized in Table 1.1.

Stationary iterative methods converge if and only if the spectral radius of the iteration matrix (i.e., the largest absolute value of its eigenvalue) is less than 1. Although the convergence rate of linear stationary iterations can be prohibitively slow in general, preconditioning can be applied for faster convergence.

Krylov Subspace Methods

Krylov subspace methods are among the most powerful and robust techniques for solving linear systems. These methods work in the Krylov subspace, i.e.,

$$\mathcal{K}_r(A, b) = \text{span}\left\{b, Ab, A^2b, \ldots, A^{r-1}b\right\}$$

built using matrix–vector products. Since a naïve application of matrix–vector products can lead to subspace vectors being ill-conditioned, an orthonormalization process, such as the Gram–Schmidt method, is typically performed in classical algorithms, which requires the computation of inner products, a nonlinear operation.

The Krylov subspace can be applied to eigenvalue problems as shown in Chap. 29: Eigenvalue Problems.

Kronecker Products

Kronecker products arise frequently in quantum computing and have some important properties:

Kronecker products are bilinear and associative:

$$A \otimes (B + C) = A \otimes B + A \otimes C$$

Kronecker products are non-commutative:

$$B \otimes A \neq A \otimes B$$

The matrices in a Kronecker product operate within their subspaces. This is evident from the mixed-product property:

$$(A \otimes B)(C \otimes D) = AC \otimes BD$$

This readily extends to eigendecompositions (if it exists for each matrix in the Kronecker product):

$$A \otimes B = \left(V_A D_A V_A^\dagger\right) \otimes \left(V_B D_B V_B^\dagger\right)$$

A sum of Kronecker products of the form

$$K_{AB} = A \otimes I_n + I_m \otimes B$$

where $A \in \mathbb{C}^{m \times m}$, $B \in \mathbb{C}^{n \times n}$ can be exponentiated as

$$\exp(K_{AB}) = \exp(A) \otimes \exp(B).$$

Polynomial Approximations

2

Polynomial approximations are frequently used in developing and analyzing quantum algorithms. Advanced quantum algorithms, especially block encoding techniques and quantum signal processing, rely on polynomial approximations of analytic matrix functions. We summarize the main approximation techniques, which fall into two categories: (1) pointwise approximations (e.g., Taylor expansions), and (2) uniform approximations over an interval (e.g., Chebyshev polynomial expansions).

Approximation Around a Point

Let $f : D \to \mathbb{C}$ be an analytic function defined on an open set $D \subseteq \mathbb{C}$, and let $a \in D$. The Taylor series of f about $x = a$ is

$$f(x) = \sum_{i=0}^{\infty} \frac{f^{(i)}(a)}{i!}(x-a)^i$$

where $f^{(i)}$ denotes the i-th derivative of f.

The degree-n Taylor polynomial, denoted as $T_n(x)$, is defined by truncating the series:

$$T_n(x) = \sum_{i=0}^{n} \frac{f^{(i)}(a)}{i!}(x-a)^i$$

The approximation error at $x \in D$ is given by the remainder term,

$$R_n(x) = f(x) - T_n(x) = \frac{f^{(n+1)}(c)}{(n+1)!}(x-a)^{n+1}$$

O. M. Raisuddin and S. De, *Quantum Computing for Engineers*,
https://doi.org/10.1007/978-3-032-03325-3_2

for some $c \in [a, x]$. The error bound is then

$$|f(x) - T_n(x)| = |R_n(x)| \leq \sup_{c \in [a,x]} \left| \frac{f^{(n+1)}(c)}{(n+1)!}(x-a)^{n+1} \right|$$

As $n \to \infty$, the Taylor series converges to $f(x)$ for all x within the radius of convergence [1].

Approximation Over an Interval

Suppose we wish to approximate f uniformly on a closed interval $[a, b]$. Without loss of generality, we may map $[a, b]$ to $[-1, 1]$ via an affine transformation. For: $[-1, 1] \to \mathbb{R}$, a common approach is to expand f in terms of Chebyshev polynomials of the first kind, $T_n(x)$), defined as

$$T_n(x) = \cos(n \arccos x),\ x \in [-1, 1],\ n \geq 0$$

These polynomials satisfy the recurrence relations:

$$T_0(x) = 1$$

$$T_1(x) = x$$

$$T_{n+1}(x) = 2xT_n(x) - T_{n-1}$$

The roots of Chebyshev polynomials of the first kind, also known as Chebyshev nodes, are the projections of $n + 1$ equally spaced points on a semicircle of unit radius onto its diameter, as shown in Fig. 2.1.

We may approximate $f(x)$ over an interval as

$$f(x) \approx P_n(x) = \sum_{i=0}^{n} \alpha_i T_i(x)$$

To approximate a function $f(x)$ in a Chebyshev basis, we first map the interval $x \in [a, b]$ to $\tilde{x} = \frac{2x-(a+b)}{b-a} \in [-1, 1]$ to obtain $f(\tilde{x})$, and then sample $f(\tilde{x})$ at $n+1$ Chebyshev nodes. One may then fit the polynomial $P_n(\tilde{x})$ of degree n to the $n+1$ points $\left\{\tilde{x}_k, f(\tilde{x}_k) | \tilde{x}_k = \cos\left(\frac{2k+1}{2n}\pi\right) \forall\ k \in [0, n] \subset \mathbb{Z}\right\}$.

The Chebyshev approximation of a function bounds the approximation error $f(x) - P_n(x)$ in the max norm over an interval, i.e., $\max_{x \in [a,b]} |f(x) - P_n(x)|$. For the interval $[a, b]$, this is

$$|f(x) - P_n(x)| \leq \frac{1}{2^n(n+1)!} \left(\frac{b-a}{2}\right)^{n+1} \max_{c \in [a,b]} \left|f^{(n+1)}(c)\right|$$

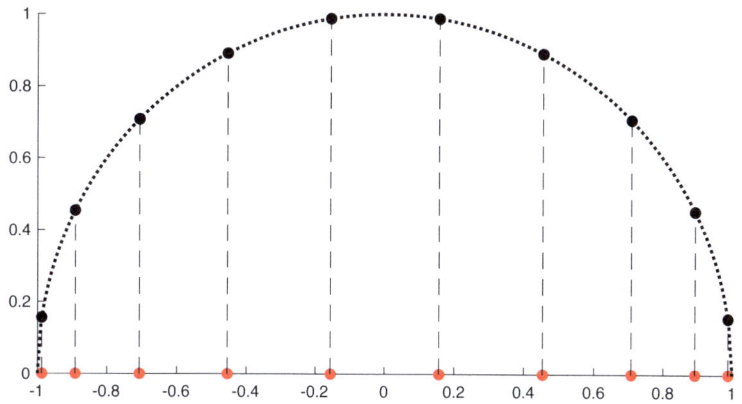

Fig. 2.1 The placement of Chebyshev nodes is visualized as equidistant points on a semicircle projected onto its diameter

The Chebyshev approximation is typically considered to yield the minimax (best uniform) polynomial approximation for continuous functions over $x \in [a, b]$, achieving the smallest possible maximum error in the interval. However, this optimality does not always hold for $x \notin [a, b]$, the Chebyshev minimax polynomial may not minimize the approximation error, and, in certain cases, alternative polynomial constructions can yield smaller errors outside the original interval [2]. Nevertheless, Chebyshev polynomials are widely used due to their robustness for analytic functions, and they can also provide good approximations for some non-analytic functions. The Remez exchange algorithm is an iterative method for constructing the minimax polynomial and can be used to compute the coefficients for the Chebyshev (and other minimax) approximations [3].

Analytic Functions of Matrices

The concept of evaluating analytical functions may be extended to square matrices using a Taylor series approximation:

$$f(A) \approx \sum_{i=0}^{n} \alpha_i A^i$$

where α_i are the coefficients of the Taylor series approximation of f.

If A is diagonalizable, one may use the diagonalization instead to compute $f(A)$ exactly as

$$\begin{aligned} f(A) &= \sum_{i=0}^{\infty} \alpha_i A^i = \sum_{i=0}^{\infty} \alpha_i \left(V \Lambda V^{-1}\right)^i = \sum_{i=0}^{\infty} \alpha_i V \Lambda^i V^{-1} \\ &= V \sum_{i=0}^{\infty} \left[\alpha_i \Lambda^i\right] V^{-1} = V f(\Lambda) V^{-1}. \end{aligned}$$

where Λ is a diagonal matrix containing the eigenvalues λ_j, and $f(\Lambda)$ is a diagonal matrix with entries $f(\lambda_j)$ on the diagonal.

Matrix Exponentiation

Exponentiation of a matrix A as e^{iA} is an important task in quantum computing since it can exactly solve time-independent systems of ordinary differential equations, e.g., the time-dependent Schrödinger equation of an isolated quantum mechanical system.

If the diagonalization of A is available, one may simply compute the matrix exponential as an analytical function of a matrix as discussed previously. However, the diagonalization of the system is typically not available, and is often intractable to compute for large N. Therefore, various approximations of matrix exponentials have been developed, of which the most notable are Trotter–Suzuki formulas, Taylor series approximations, and quantum signal processing techniques.

The problem of approximating a matrix exponential of the form e^{iH} (or the operation of this matrix exponential on a ket $|\psi\rangle$) where H is a Hamiltonian is known as a Hamiltonian simulation problem (for isolated systems). We deal with this topic in detail in Chap. 28: Hamiltonian Simulation Techniques. Here, we simply provide a mathematical introduction to the problem.

Typically, H is a sparse matrix which can be expressed as a sum of 1-sparse terms H_i:

$$H = \sum_i H_i$$

and each H_i can be exponentiated individually in a straightforward manner. Exponentiating these 1-sparse terms can be done easily on a quantum computer.

However, the fundamental problem arises in combining these individual terms. While for $x, y \in \mathbb{C}$, $e^{x+y} = e^x e^y$, this holds for matrix exponentials if and only if $A, B \in \mathbb{C}^N$ commute, i.e., if

$$[A, B] = AB - BA = 0$$

is true, only then $e^{A+B} = e^A e^B$. In most problems of practical interest, H cannot be expressed as a sum of commuting 1-sparse terms.

For the general case of arbitrary A and B, the Baker–Campbell–Hausdorff (BCH) formula provides the exact result:

$$e^X e^Y = e^Z$$

where

$$Z = X + Y + \frac{1}{2}[X, Y] + \frac{1}{12}[X, [X, Y]] - \frac{1}{12}[Y, [X, Y]] + \ldots$$

The Lie–Trotter (and Suzuki–Trotter) approximation [4] is one of the most well-known and simplest methods for approximating a matrix exponential on quantum computers, for which the BCH formula provides error bounds. Details of Trotter–Suzuki formulas are provided in Chap. 21: Trotterization. The Trotter–Suzuki method belongs to a class of techniques known as "product formulas," and newer product formulas have also been discovered with improved error bounds [5–8].

References

1. I. Stewart, D. Tall, *Complex Analysis*, 1st ed. (Cambridge University Press, 1983). https://doi.org/10.1017/CBO9781139171632
2. B. Fischer, R. Freund, Chebyshev polynomials are not always optimal. J. Approx. Theory **65**(3), 261–272 (1991). https://doi.org/10.1016/0021-9045(91)90091-N
3. E. Remez, Sur le calcul effectif des polynomes d'approximation de tchebichef. CR Acad. Sci. Paris (1934)
4. M. Suzuki, Generalized Trotter's formula and systematic approximants of exponential operators and inner derivations with applications to many-body problems. Commun. Math. Phys. **51**(2), 183–190 (1976). https://doi.org/10.1007/BF01609348
5. C.-H. Cho, D.W. Berry, M.-H. Hsieh, Doubling the order of approximation via the randomized product formula. Phys. Rev. A **109**(6), 062431 (2024). https://doi.org/10.1103/PhysRevA.109.062431
6. M.E.S. Morales, P.C.S. Costa, D.K. Burgarth, Y.R. Sanders, D.W. Berry, Greatly improved higher-order product formulae for quantum simulation (2022). https://doi.org/10.48550/ARXIV.2210.15817
7. M.E.S. Morales, P.C.S. Costa, G. Pantaleoni, D.K. Burgarth, Y.R. Sanders, D.W. Berry, Selection and improvement of product formulae for best performance of quantum simulation (2025). https://doi.org/10.48550/arXiv.2210.15817
8. M. Bagherimehrab et al. Faster algorithmic quantum and classical simulations by corrected product formulas (2025). https://doi.org/10.48550/arXiv.2409.08265

3 Theory of Computing

In this chapter, we provide an overview of the theoretical foundations of computing, and, in the next chapter, we will present how these foundations come to fruition as modern electronic computers.

Automata and Turing Machines

We will begin our discussion with various types of *automata*, an abstract object that follows a sequence of instructions. An automaton may be built as a machine or could be a human following instructions in a "desultory manner." One may establish a set of rules to describe various types of automata and study the limits of their computation. Ideally, we would like a device to be able to follow any sequence of instructions to compute any function that we can compute algorithmically. We will briefly study this problem by establishing a few automata and endowing them with properties until they are powerful enough to compute any function that has an algorithm to compute it.

Perhaps the simplest form of an automaton is *Combinational Logic*. This can often be expressed as a Boolean expression or a Boolean Circuit as shown in Fig. 3.1.

The behavior of this automaton (the output) depends solely on the current inputs, regardless of any previous inputs. The limitations of a Combinatorial Logic automaton are quite obvious. Consider a Combinatorial Logic Automaton that computes the parity of a binary string of length n. To compute the parity of a binary string s of length $2n$ bits with such an automaton, one may "remember" the result of the first and last n bits, and compute the parity as $parity(parity(s[0 : n - 1]), parity(s[n : 2n - 1]))$. However, this automaton is incapable of any form of inference beyond its immediate input; it does not have

O. M. Raisuddin and S. De, *Quantum Computing for Engineers*,
https://doi.org/10.1007/978-3-032-03325-3_3

Fig. 3.1 A Combinational Logic statement and its equivalent logic circuit

$$D = (A \wedge B) \vee C$$

any memory or a state dependent on a previous computation. Therefore, we cannot compute the parity of arbitrarily long strings solely with this automaton.

However, it is imperative to note that for limited input and output sizes n and m respectively, there always exists a corresponding circuit for any arbitrary function $f : \{0, 1\}^n \to \{0, 1\}^m$. Toward the end of this chapter, we will revisit this fact.

One may form a more complex device by allowing the device to assume a finite number of *states*. This class of devices is known as *finite state automata*. These devices may assume a finite number of states and transition between them based on the current state and current input. They do not have a memory of previous states or how they arrived at their current state.

Let's describe a finite state automaton to compute the parity of strings. Our automaton can assume two states. q_{even} and q_{odd}, corresponding to an even or odd parity. The automaton is initialized in the state q_{even}. The automaton accepts a binary bit $x = 0$ or $x = 1$, one at a time, as input. Whenever a binary $x = 1$ is input to the automaton, it will change its state from q_{even} to q_{odd} and vice versa. Whenever a binary $x = 0$ is input to the automaton, it will not change its state. This is summarized in Fig. 3.2, which describes the state diagram and transition function $\delta : q, x \to q$. After initializing the automaton in the state q_{even}, all the bits of a string are fed to the automaton one by one, and, at the end, the state of the automaton indicates the parity of the bitstring.

By simply equipping our device with a finite number of states, we have expanded its computational capability: a finite state automaton can compute the parity of an arbitrarily long binary string. But can a finite state automaton check whether an arbitrary binary string is balanced, i.e., whether the number of ones

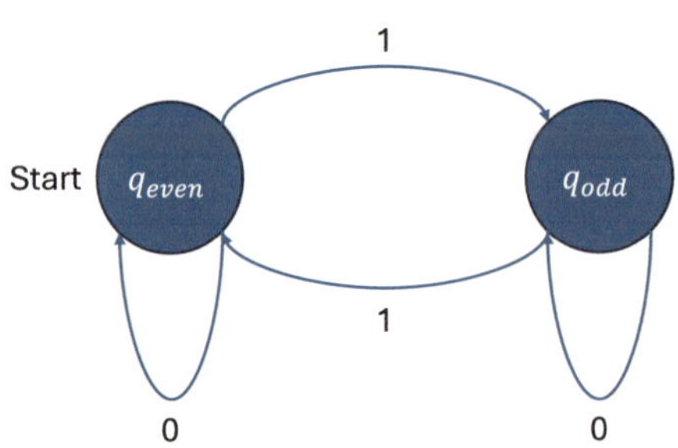

q	x	$\delta(q, x)$
q_{even}	1	q_{odd}
q_{even}	0	q_{even}
q_{odd}	1	q_{even}
q_{odd}	0	q_{odd}

Fig. 3.2 A finite state automaton for computing the parity of a string of bits

matches the number of zeros, or not with this form of automata? This is not possible, since this requires keeping track of the difference between the number of ones and zeros encountered in the string, and for an arbitrarily long string, this can require an infinite number of states to keep track of this difference! We have encountered a limitation of our automaton once again.

We will now equip our automata with an idealized infinite memory. We are now arriving at a description of a deterministic Turing Machine [1], which we will simply refer to as a Turing machine for brevity. There are many subtly different descriptions of Turing machines, all of which are similar and equally powerful. Here, we will provide a description similar to [2] for intertextual convenience and recommend the description of automata in [3] for readers interested in thoroughness.

A Turing machine consists of the following elements, as visualized in Fig. 3.3:

- Tape
- Read–write tape head
- Finite state control
- Program.

The tape in a Turing machine is infinitely long in one direction and is divided into squares. In the following, we adopt the convention that the tape stretches infinitely to the right. Each square can store one symbol from a set of possible symbols known as an *alphabet* Γ.

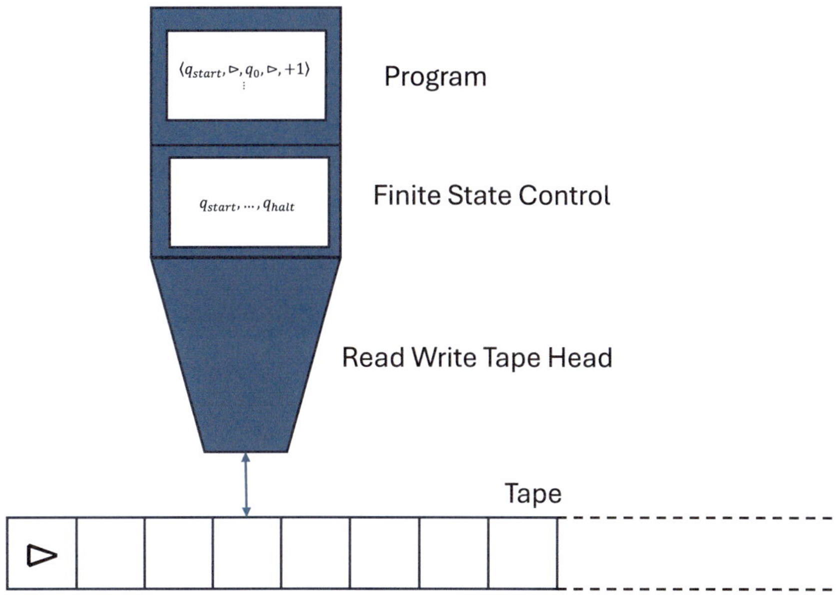

Fig. 3.3 Schematic depiction of a Turing machine

At every step, the read–write tape head is positioned over a particular square. When the Turing machine is initialized, the read–write tape head is positioned over the leftmost square, and all information needed by the computation is written on a finite number of the leftmost squares of the tape, and the remainder of the tape is left blank. In each step, the read–write tape head reads the current square it is positioned on and

1. writes the current square: $\Gamma \rightarrow \Gamma$
2. moves the head one square to the left or right, or remains stationary. This is described by the set of movements $S : \{-1, +1, 0\}$.

The finite state control describes an internal state $q \in Q$ of the machine (this does not include the square being read by the tape head), and the states are labeled as $q_i \in Q$ where the cardinality $|Q| = m$ where m is sufficiently large and finite. There are two special states q_s and q_h, the start state and halt state, respectively. When the machine is initialized, the finite state control is in the state q_s. When the machine enters a halt state q_h (there may be multiple halt states), it ceases operation.

Note that a Turing machine may loop forever! We will return to this issue later in the chapter. When a Turing machine does halt, to indicate the reason for the termination of a program it is important to indicate two subsets of Q: the *accept* states $q_a \in Q_a \subset Q$ and *reject* states $q_r \in Q_r \subset Q$ s.t. $Q_a \cap Q_r = \emptyset$. A Turing machine will halt on these states. Using this terminology, a Turing machine can *decide* an input, i.e., accept or reject it.

The *program* of a Turing machine is a list of *program lines*. To avoid any confusion, we emphasize that, unlike typical modern computer programs we are familiar with, the sequence in which the program lines are presented is not necessarily the sequence of steps the Turing machine will step through. Instead, the program is better described as a transition function δ : Q, $\Gamma \rightarrow$ Q$\prime$, Γ', S and each program line is a quintuple q, x, q', x', s, where

- $q \in Q$ is the current state of the machine
- $x \in \Gamma$ is the current symbol read by the read–write tape head
- $q' \in Q$ is the next state of the machine
- $x' \in \Gamma$ is the symbol to be written in the current square
- $s \in S$ is the direction in which to move the read–write head.

Execution of any program line is counted as one computational step (regardless of the lookup through all the program lines). If the machine encounters a state and tape square q, x for which it cannot match a program line, it will automatically enter the halting state q_h. If the read–write head is on the leftmost square and is instructed to move further to the left, it will not move.

The automaton we have described above may seem deceptively simple, but it captures the essence of all classical computation. Any function whose computation can be described as an algorithm can be computed by a Turing machine. More generally, a Turing machine can execute any algorithm. This is summarized in the Church–Turing thesis:

A Turing machine can simulate any algorithmic process.
Let's return to our problem of determining whether a binary string is balanced or not. In the context of Turing machines, this problem is often referred to as a balanced parenthesis problem. We will use a Turing machine with the following description:

Alphabet: $\Gamma : \{0, 1, b, c, \triangleright\}$

- binary 1
- binary 0
- blank b
- "canceled" square c
- leftmost square marker $\triangleright$.

Tape: Leftmost square is initialized as $\triangleright$, followed by the string as consecutive zeros and ones. The remainder of the tape is blank.

Finite state control:
q_{start}: machine initial state
q_{scan}: scan right for 0 or 1 till blank is encountered
q_{return}: return to leftmost square
q_{c0}: scan right till a 1 is found to cancel a 0
q_{c1}: scan right till a 0 is found to cancel a 1
q_{bal}: accept state, string is balanced
q_{unbal}: reject state, string is unbalanced.

Program:

Program Line	Description	Task
$q_{start}, \triangleright, q_{scan}, \triangleright, +1$	Start	–
$q_{scan}, 0, q_{c0}, c, +1$	Found 0, enter cancel 0 state	Scan toward right till a 0 or 1 is encountered and enter corresponding cancel state. If end of string reached, accept as balanced
$q_{scan}, 1, q_{c1}, c, +1$	Found 1, enter cancel 1 state	
$q_{scan}, c, q_{scan}, c, +1$	Already canceled, continue	
$q_{scan}, b, q_{bal}, b, 0$	All bits matched, accept	

(continued)

(continued)

Program Line	Description	Task
$q_{return}, \triangleright, q_{scan}, \triangleright, +1$	Reached first square, start scanning	Return toward left till first square, then enter scan state
$q_{return}, c, q_{return}, c, -1$	Continue toward left	
$q_{return}, 0, q_{return}, 0, -1$	Continue toward left	
$q_{return}, 1, q_{return}, 1, -1$	Continue toward left	
$q_{c0}, 1, q_{return}, c, 0$	Cancel 0 with 1	Move right till a 1 is encountered to cancel a 0. If end of string is reached and 1 is not found, reject as unbalanced
$q_{c0}, c, q_{c0}, c, +1$	Already canceled, continue	
$q_{c0}, 0, q_{c0}, 0, +1$	0, continue	
$q_{c0}, b, q_{unbal}, b, 0$	Could not cancel 0 with 1, reject	
$q_{c1}, 0, q_{return}, c, 0$	Cancel 1 with 0	Vice versa for above
$q_{c1}, c, q_{c1}, c, +1$	Already canceled, continue	
$q_{c1}, 1, q_{c1}, 1, +1$	1, continue	
$q_{c1}, b, q_{unbal}, b, 0$	Could not cancel 1 with 0, reject	

We now see that our automata can determine whether strings are balanced or not. However, is there a limit to what Turing machines can compute? As long as there is an algorithm with a finite number of steps, a Turing machine is capable of computing it.

Several uncomputable problems have been discovered. The first and most famous counterexample of a problem which a Turing machine (or any known automaton) cannot solve is the halting problem, since an algorithm to solve it does not and cannot exist. In fact, Alan Turing devised the concept of Turing machines to answer a specific problem posed by David Hilbert, the *Entscheidungsproblem* or "decision" problem. Loosely defined, the decision problem questions:

Does there exist a general algorithm that van take as input a (logical, mathematical, etc.) statement and "decide" whether it is valid or not?

To the Editor,
The Computer Journal.

An impossible program

Sir,
A well-known piece of folk-lore among programmers holds that it is impossible to write a program which can examine any other program and tell, in every case, if it will terminate or get into a closed loop when it is run. I have never actually seen a proof of this in print, and though Alan Turing once gave me a verbal proof (in a railway carriage on the way to a Conference at the NPL in 1953), I unfortunately and promptly forgot the details. This left me with an uneasy feeling that the proof must be long or complicated, but in fact it is so short and simple that it may be of interest to casual readers. The version below uses CPL, but not in any essential way.

Suppose $T[R]$ is a Boolean function taking a routine (or program) R with no formal or free variables as its argument and that for all R, $T[R] =$ **True** if R terminates if run and that $T[R] =$ **False** if R does not terminate. Consider the routine P defined as follows

rec routine P

§L : **if** $T[P]$ **go to** L

Return §

If $T[P] =$ **True** the routine P will loop, and it will only terminate if $T[P] =$ **False**. In each case $T[P]$ has exactly the wrong value, and this contradiction shows that the function T cannot exist.

Yours faithfully,
C. Strachey.

Churchill College,
Cambridge.

Fig. 3.4 A proof of the halting problem [4]

Alan Turing, in his paper "On Computable Numbers, with an Application to the Entscheidungsproblem" [1], showed that a Turing machine can execute any arbitrary algorithm, but is incapable of solving the *halting problem*, answering the Entscheidungsproblem in the negative. The halting problem can be stated informally as

Given a description of a Turing machine A, can Another turing machine B take as input the description of A and decide if it halts (does not loop forever)?

There are various proofs in the negative: such a general algorithm *cannot exist* for an arbitrary input and input size. We provide here an exceedingly succinct proof in its original form [4] in Fig. 3.4.

Turing machines (and its equivalents and variants) are the most powerful known computational devices and can execute any algorithm given enough time.

Variants of Turing Machines

For completeness, we will now briefly go over some variants of Turing machines, many of which are equivalent to the machine described above.

Although Turing machines in their simplicity are quite powerful, writing programs for them can be quite laborious. To alleviate this arduousness, one may equip the Turing machines with multiple tapes. It is straightforward to show that a multi-tape Turing machine is at least as powerful as a single tape Turing machine (simply use only one tape). Furthermore, it can also be shown that a single tape Turing machine can efficiently (polynomially equivalent) "simulate" a multi-tape Turing machine as follows.

Theorem ([5]) Given any k-string Turing machine M operating within time $f(n)$, we can construct a Turing machine M' operating within time $O\big(f(n)^2\big)$ and such that, for any input x, $M(x) = M'(x)$.

Table 3.1 A summary of various automata

	Combinatorial logic	Finite state automata	Deterministic Turing machine
Computation depends on	Inputs	Inputs and current state	Inputs, current state, and current tape-square state
Cannot compute	Parity of arbitrary binary strings with unlimited input size	Balance of an arbitrarily long bitstring	Anything for which an algorithm does not exist (e.g., halting problem)

Furthermore, to reconcile the differences between various flavors of Turing machines with varying alphabets, numbers and types of tapes, internal states, and programs, a *Universal Turing Machine* can be formulated. A Universal Turing Machine has the description of the Turing Machine it is simulating written on its tape, including its program, which is evocative of the operation of modern classical computers which have their programs (or even a virtual machine) written out in memory (although modern classical computers are better described by RAM machines, which we do not discuss in this book).

We now summarize the various automata we have encountered and their salient features in Table 3.1.

The Turing machines we have described and discussed above are *deterministic* in nature, whose program lines implement a transition function of the form $\delta : (q, x) \to (q, x, s)$. One may also study *non-deterministic* Turing machines, whose program is written as a set of outcomes $\delta : (q, x) \to \{(q, x, s)\}$. Note that a probability is not assigned to these outcomes; rather, the Turing machine can "explore" all branches of a computational tree at once. We emphasize that such a machine is not known to be constructible, rather it is a theoretical tool to study the complexity of computation. Non-deterministic Turing machines are not more powerful than deterministic Turing machines. One may always describe a deterministic Turing machine that exhaustively explores all computational branches. The difference lies in *efficiency*, i.e., the resources and time needed to solve a problem. These concepts will be explored in Chap. 5: Information and Complexity Theory. Quantum computers are *not equivalent* to non-deterministic Turing machines. There is no known physical instantiation of a non-deterministic Turing machine, and it is widely accepted to be impossible to create one.

We note that if probabilities are assigned to the set of outcomes of a non-deterministic Turing machine, and the resulting machine occupies a *unique* configuration at each step (rather than exploring every branch at once), it is a *probabilistic Turing machine*. Probabilistic Turing machines can be more efficient than deterministic Turing machines, but are less efficient than non-deterministic Turing machines. To set the stage for the later chapters, we will now briefly go over the idea of mapping the operation of a Turing machine to Boolean circuits.

Universal Circuit Families

While the theoretical construct of Turing machines is powerful and allows the study of problems in terms of their decidability and complexity, building Turing machines with infinite memory is impossible, and building Turing machines with finite memory is impractical, in contrast to Boolean circuits. However, at the beginning of this chapter, we concluded that

(i) Boolean circuits can compute any function $f : \{0, 1\}^m \rightarrow \{0, 1\}^n$
(ii) For the parity problem of size $\tilde{m}$, we cannot use a single Boolean circuit that computes $parity : \{0, 1\}^m \rightarrow \{0, 1\}^1$ where $\tilde{m} > m$.

Naively seeking an equivalence between Turing machines and Boolean circuits quickly leads to contradictions. For example, since Boolean circuits can compute any $f : \{0, 1\}^m \rightarrow \{0, 1\}^n$, a Boolean circuit does exist for each m which solves the halting problem for an input size m. But we know that the halting problem is undecidable.

The other difference is that the same Turing machine (with its program lines, etc.) can be used to compute problems of arbitrary input size, e.g., the length of a string whose parity is to be determined. However, Boolean circuits do not satisfy this property for many problems. For this reason, Turing machines are a *uniform* model of computation, and Boolean circuits are a *non-uniform* model of computation.

We will now reconcile the differences between Turing machines and Boolean circuits by considering a subset of all possible circuits: the *uniform circuit family* denoted by $\{C_m\}$. The uniform circuit family consists of circuits such that, given an input size m for a problem, a Turing machine can output a description of an equivalent Boolean circuit that computes $f : \{0, 1\}^m \rightarrow \{0, 1\}^n$. Turing machines and uniform circuit families compute the same class of functions, i.e., there must exist an algorithm to compute f. Note that since an algorithm for the halting problem is known not to exist, uniform circuit families exclude such circuits.

With this restriction on $\{C_m\}$, we can safely say that while Boolean circuits may exist for various m which may decide the halting problem, we do not (cannot) have an algorithm to prepare them and therefore these circuits do not lie in $\{C_m\}$.

As we will see in the next chapter, uniform circuit families are not exactly how modern classical computers are built and programmed. However, the quantum circuit computational model, introduced in Chap. 10: Quantum Measurements and Circuits, is by far the most convenient and popular method to program quantum computers.

Finally, we note that modern classical computers, given infinite/sufficient memory and byte length, are computationally as powerful as Turing machines (they can readily simulate a Turing machine). However, classical computers also possess the capability to randomly access any data in memory, unlike Turing machines, which move one square at a time. Therefore, modern classical computers are closer to

RAM machines from an architectural perspective, which are polynomially equivalent to Turing machines. Even so, compute operations are executed in a CPU by combinational circuits (or a sequence of combinational circuits).

In this chapter, we have explored the theoretical foundations of computing by examining various automata, their computational power, and the limits of computability in the context of algorithms. In the next chapter, we will go over how classical computers are implemented as the practical counterpart to these theoretical foundations.

References

1. A.M. Turing, On Computable Numbers, with an Application to the Entscheidungsproblem. *Proceedings of the London Mathematical Society*, vol. s2–42, no. 1, pp. 230–265 (1937). https://doi.org/10.1112/plms/s2-42.1.230
2. M.A. Nielsen, I.L. Chuang, *Quantum Computation and Quantum Information: 10th Anniversary Edition*, 1st ed. (Cambridge University Press, 2012). https://doi.org/10.1017/CBO9780511976667
3. R. Manenti, M. Motta, *Quantum Information Science* (Oxford University Press, Incorporated, Oxford, 2023)
4. Strachey, An impossible program. Comput. J. **7**(4), 313–313 (1965). https://doi.org/10.1093/comjnl/7.4.313
5 C.H. Papadimitriou, *Computational complexity* (Addison-Wesley, Reading (Mass), 1994)

4 An Overview of Practical Classical Computing

We will now turn our attention to the practical implementations of computers. An elementary component of classical computing is logic gates connected by wires. Some logic gates that may be familiar to the reader are NOT, AND, and OR (Fig. 4.1, Tables 4.1 and 4.2). For completeness, we will also need to include the FANOUT and CROSSOVER operations, which split a single wire into multiple wires and allow wires to cross over each other, respectively, without affecting the information they carry [1]. What is particularly important about these gates is that, together with the FANOUT and CROSSOVER operations, they form a classical *universal gate set* (not the same as a universal circuit family). A classical universal gate set can represent any arbitrary Boolean operation on bits. We investigated *combinatorial* Boolean circuits as Combinatorial Logic automata in the previous chapter.

Another universal gate set consists solely of the NAND gate [2]. Therefore, any classical Boolean operation can be represented using NAND gates. Besides being optimally small (only 1 gate to form a universal gate set), it requires fewer transistors to implement and is therefore often preferred for manufacturing Very Large-Scale Integration (VLSI) circuits. This chapter is not geared toward designing our own state-of-the-art computer, so we shall not restrict the following discussion to the NAND gate.

Transistors as Physical Logic Gates

Transistors, vacuum tubes, and even fluid valves can be used to construct physical logic gates. As an example, Fig. 4.2 shows a layout of an AND gate constructed using two NPN (NPN indicates the doping type of the silicon) bipolar junction transistors and three electrical resistors.

O. M. Raisuddin and S. De, *Quantum Computing for Engineers*,
https://doi.org/10.1007/978-3-032-03325-3_4

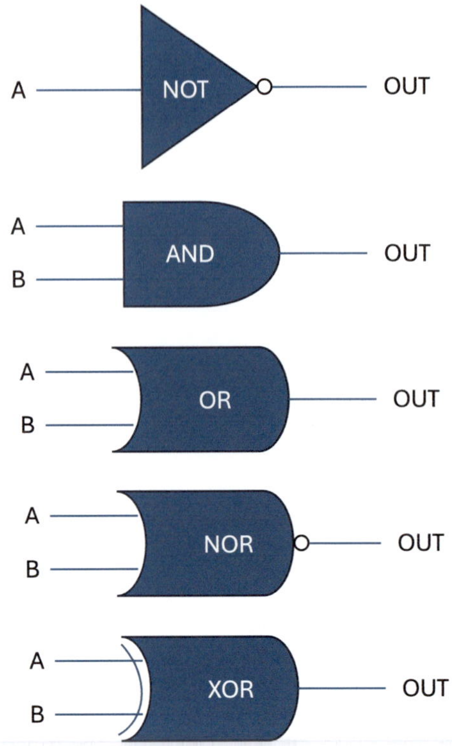

Fig. 4.1 Common Boolean logic gates and their symbols

Table 4.1 Logical truth table for AND and OR Boolean gates

		OUT	
A	B	AND	OR
1	1	1	1
0	1	0	1
1	0	0	1
0	0	0	0

Table 4.2 Logical truth table for a Boolean NOT gate

	OUT
A	NOT
0	1
1	0

The voltage levels of the pins A, B, and OUT determine their logical state. The voltage levels assigned to Logical 1 and Logical 0 depend on the type of transistor technology underlying the gate. Some examples are given in Table 4.3. However, we can abstract away these details by simply considering logical states of 0 and 1.

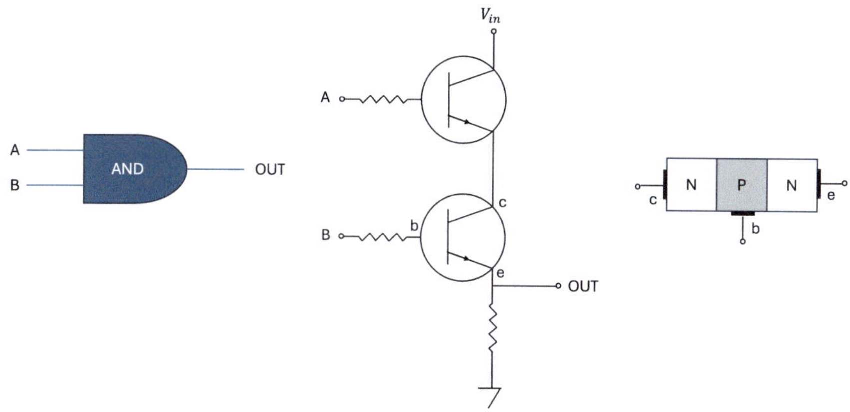

Fig. 4.2 Left: AND gate. Center: AND gate implemented using bipolar junction transistors of NPN type. Right: Schematic of a bipolar junction transistor of NPN type with labeled doped regions

Table 4.3 Logical voltage ranges for various types of transistors

	Transistor type	
Logical state	TTL	CMOS
1	2.7–5.0 Volts	2.4–3.3 Volts
0	0.0–0.4 Volts	0.0–0.5 Volts

By combining several gates, one may implement circuits for Boolean operations of practical interest involving multiple bits. As an example, a circuit to add two bits is given in Fig. 4.3, followed by a circuit to add two unsigned integers in Fig. 4.4.

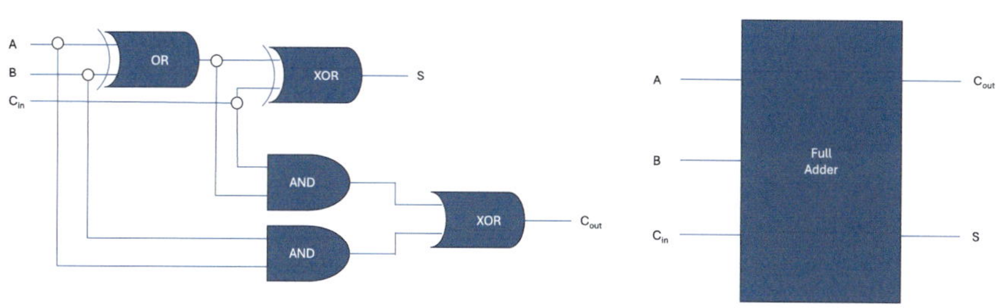

Fig. 4.3 Left: Digital circuit of a full adder. Right: Schematic of a full adder. Both diagrams show the input bits A, B with an optional "Carry in" bit for addition, and the output bit S with a "Carry out" bit

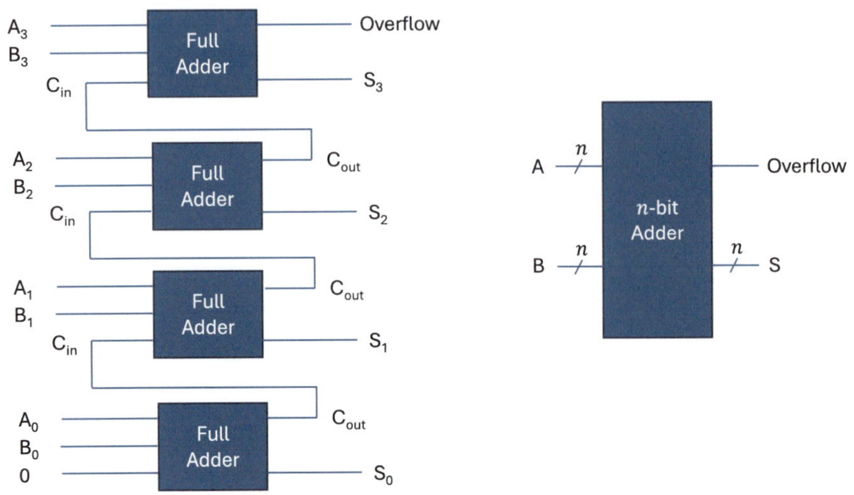

Fig. 4.4 Left: A circuit for adding two 4-bit unsigned integers with an overflow flag. Right: n-bit adder module with an overflow flag

Combinational Circuits

In the remainder of this chapter, we will keep considering this problem of adding integers and will ignore the issue of integer overflow for simplicity.

One thing to note is that these circuits are *acyclic* and at a *steady state*, i.e., the output of any gate does not impact its input, and the electrical signal has propagated through all the transistors and stabilized. This property is important for the *computation* of Boolean functions using logic gates.

This idea of forming acyclic circuits can be extended to form logical circuits for any arbitrary Boolean function, e.g., multiplication of two integers. However, continuing with this approach would require forming longer circuits for longer programs, and rewiring them for every program one would like to execute. The deeper the circuits are, the longer it takes for the signal to propagate through the entire sequence of gates, and the output is only valid when the circuit has reached a stable state; transient states are not valid in this framework! Furthermore, this approach does not allow reuse of circuits, e.g., if one would like to add N integers, it will require $\mathcal{O}(N)$ adder circuits, as illustrated for the two approaches shown in Figs. 4.5 and 4.6.

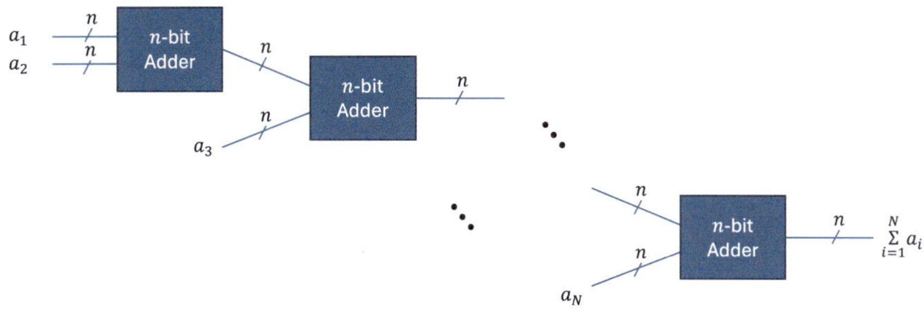

Fig. 4.5 Adding N integers using a serial addition approach with an overall circuit depth of $O(N)$

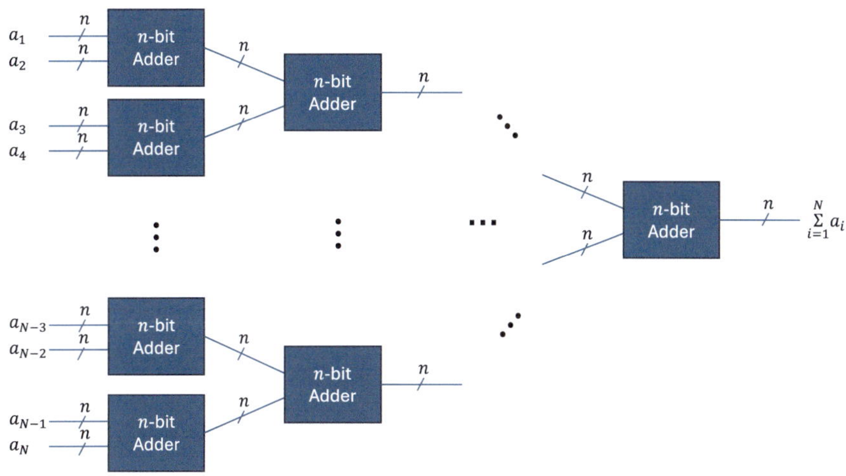

Fig. 4.6 Adding N integers using a parallel reduction approach with an overall circuit depth of $O(\log N)$

Sequential Circuits

The acyclic steady-state combinational circuit approach is not scalable and certainly not used in modern computers, which regularly crunch information on scales of Gigabytes to Exabytes. We will now bend the rules of acyclic steady-state circuits for practical purposes. A more modular approach (which will allow reuse of circuits) can be pursued using some additional elements: logical *memory* and a *clock* signal. Using these, we will now lay the groundwork for *sequential* Boolean circuits. Modern computers are manifested as a mix of sequential and combinatorial Boolean circuits.

The schematics for the components in what follows (e.g., memory cells) are presented in a way to enhance clarity and understanding. Modern hardware implements these components in a more robust and efficient manner.

Before starting, we note that if logic gates are needed for purposes of efficiency or performance, e.g., repeatedly performing the same Boolean operations for a stream of different inputs, one may implement a logic gate circuit of interest directly using either a Field-Programmable Gate Array (FPGA) or manufacture their own Application-Specific Integrated Circuit (ASIC).

A clock signal is used to synchronize operations. A typical clock signal in modern processors typically has an oscillation period of $\approx$ 4 GHz and has an idealized waveform of the form shown in Fig. 4.7.

Operations are typically "triggered" by the rising or falling edge of the clock signal. This is achieved using an "edge detector" circuit, which exploits the propagation delay of electric signals through logic gates as shown in Fig. 4.8. These elements no longer obey the steady-state assumption for Boolean circuits.

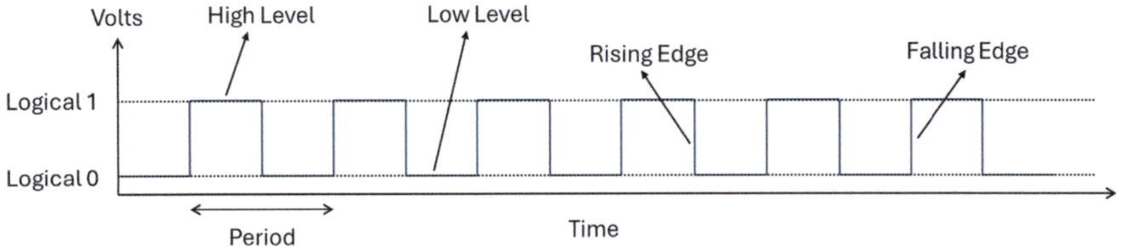

Fig. 4.7 Clock signal and associated terminology

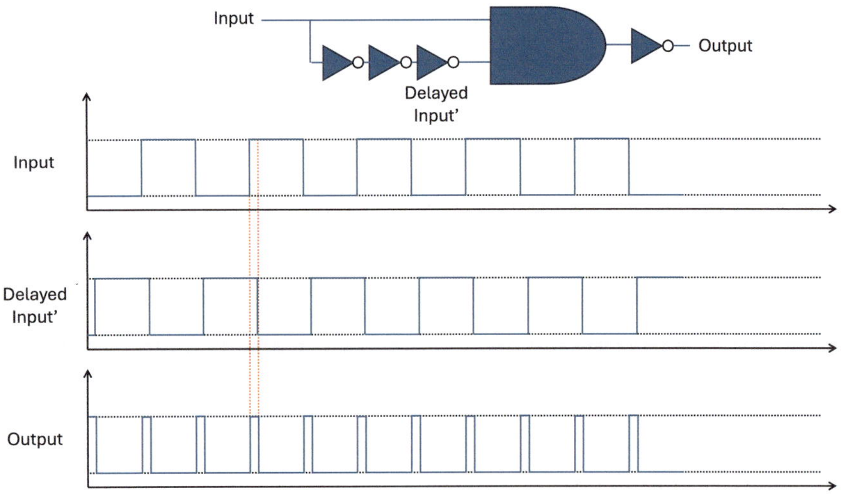

Fig. 4.8 Top: A simple edge detector circuit exploiting the delay in signal propagation through logical gates. Bottom: Input clock signal, complement of the delayed input clock signal, and the output signal spiking at the rising edge of the clock

Memory Elements

Memory elements can be constructed with the same ingredients: logic gates. A rudimentary implementation of a memory element is a "latch."

Let's investigate a simple memory element for a single bit: a Data (D) Latch, as shown in Fig. 4.9 with its logic Table 9.1. One property immediately recognizable is that this logic circuit is *cyclic*, i.e., the output of a logic gate influences its input. The inputs to the latch are Data and Enable, and the output is Q and its complement Q'. When Enable (E) pin is set to logical 0, the latch does not change its output regardless of the Data pin. However, when Enable is set to 1, the output "latches" onto the Data input state and retains it when Enable is reset to 0, i.e., it has *memory* of its state when Enable was set to 1 (Table 4.4).

We would like to be able to both *read* and *write* this single bit of memory when we *address* it. This can be achieved by adding a few more gates and a tri-state buffer to the mix. A tri-state buffer is not a logical gate, acting as an "electronic switch" instead, as illustrated in Fig. 4.10. It effectively disconnects (connects) the input and output pins when the enable pin is set to 0 (1) through a very high (low) resistance and can be implemented using bipolar junction transistors and resistors.

We can now form a *memory cell* as shown in Fig. 4.11. When Address is set to 1, the bit can be either written (Write/Read' = 1) or read (Write/Read' = 0) using the same Data pin. Otherwise, the memory cell simply stores the bit and is disconnected from the Data pin.

By putting together, say 8, such memory cells, one may form a single 8-bit (or 1 byte) memory *register* which can be read and written one byte at a time through a data bus, as shown in Fig. 4.12.

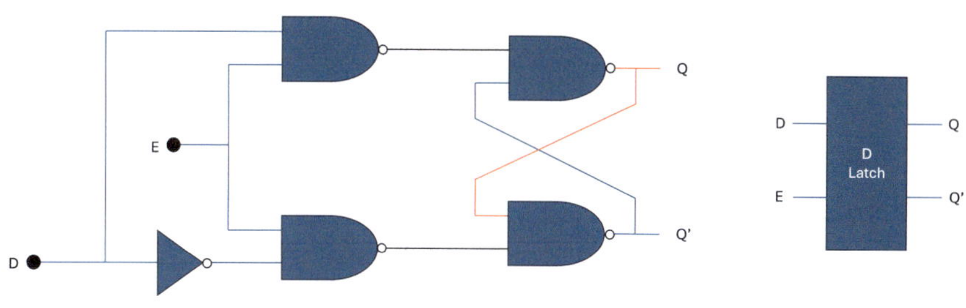

Fig. 4.9 A circuit implementing a D-Latch

Table 4.4 Logic table for a D-Latch

E	D	Q	Q'
0	*	Latch	Latch'
1	0	0	1
1	1	1	0

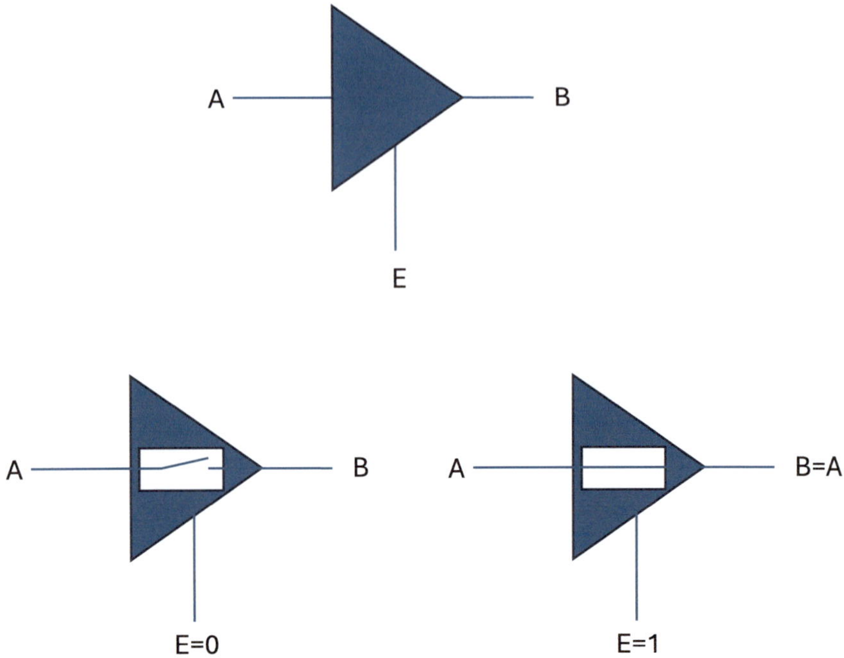

Fig. 4.10 Top: Schematic of a tri-state buffer. Bottom: Operation modes of a tri-state buffer

Furthermore, one may put together a grid of $m \times m$ memory registers to form *random-access memory* (RAM, more specifically, Static RAM) capable of storing $8 \times m^2$ bits of data using m^2 memory addresses. Figure 4.13 shows how such a rudimentary RAM can be formed using encoders and decoders.

A decoder simply "decodes" an n-bit binary string into a 2^n bit "one-hot" encoding (e.g., a binary number $b_{n-1} \ldots b_1 b_0$ is equivalent to the integer $i = \sum_{j=0}^{n-1} b_j$, and the corresponding one-hot encoding will be a binary string of length 2^n full of zeros except the $i + 1\,th$ position, i.e., b0 $\times$ 101 $= 5 \rightarrow$ 00100000). An example for two encoded bits being decoded is shown in Table 4.5.

This object can be abstracted into a RAM object with an address bus, a memory bus, and a pin for Write/Read', all synchronized by a clock. This RAM can be read one byte (8 bits) at a time by addressing a particular register.

We can organize memory into chunks to store program instructions, data, a stack, a heap, and any other information or intermediate storage needed by the program, with all memory being accessible through a memory bus consisting of an address, data, and control bus as illustrated in Fig. 4.14.

This approach of using a single memory to store both the program and data is known as Von Neumann architecture (in contrast with the Harvard architecture, which stores the program in a separate memory).

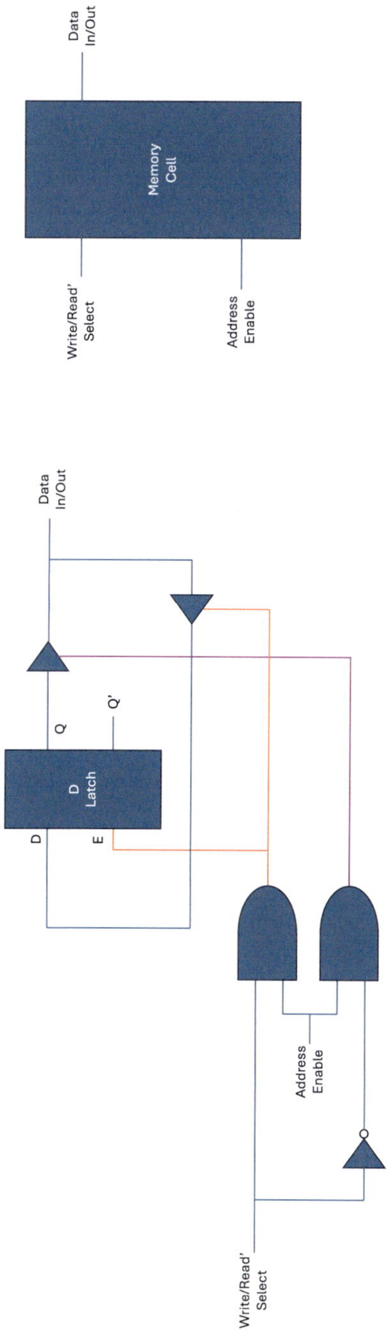

Fig. 4.11 Left: Circuit to form a memory cell from a D-Latch which can be read and written when addressed. Right: A memory cell

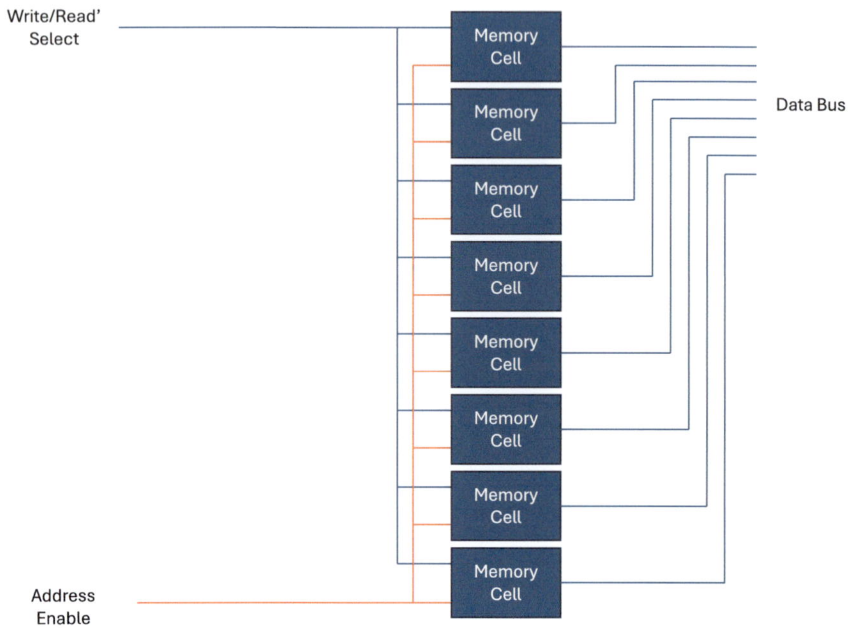

Fig. 4.12 A classical memory register with a data bus

CPU Architecture

We now turn our attention to a CPU. In its simplest form, a CPU is connected to memory and receives a clock signal. A basic CPU has a very limited number of internal registers to keep track of the CPU state and will frequently need to offload information to memory as it steps through a program. The program (or sequence of instructions) itself will be stored in memory. The CPU will "fetch" these instructions one at a time to "decode" and "execute" them (Figs. 4.15 and 4.16).

A CPU operates in a Fetch–Decode–Execute cycle:

- Fetch: Get the next instruction from memory (using the memory address of the next instruction in the program counter register).
- Decode: Write instruction to Current Instruction register to activate the appropriate circuits for the instruction (e.g., an adder circuit with inputs connected to read two general-purpose registers and output connected to write to another register)
- Execute: Execute instruction (can include read/write operations from/to memory).

Some important components of a CPU are as follows:

Registers:

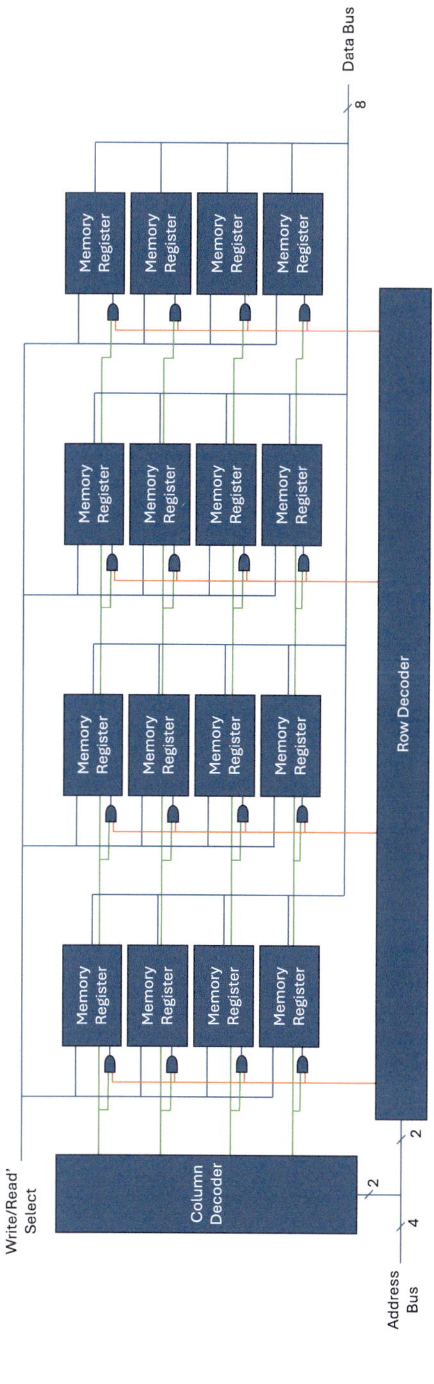

Fig. 4.13 A 4 × 4 (16) byte RAM with an address, data, and control bus

Table 4.5 Equivalent encoded and decoded bitstrings

Encoded	Decoded
00	0001
01	0010
10	0100
11	1000

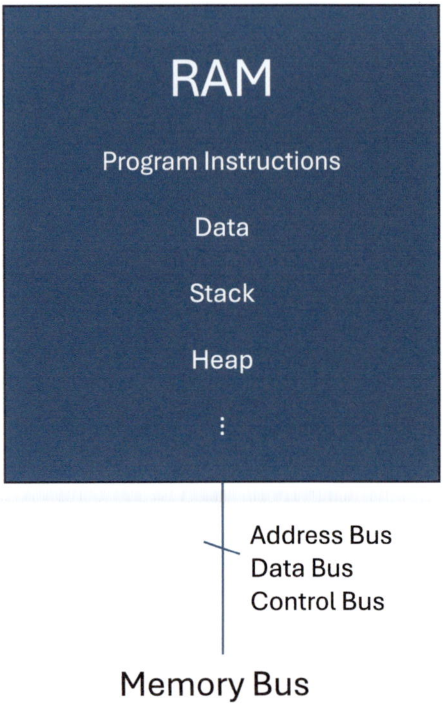

Fig. 4.14 RAM connected to a Memory Bus

- Current Instruction register: The instruction the CPU is handling in the current Fetch–Decode–Execute cycle.
- Program Counter register: Memory address of the next instruction.
- General-Purpose registers: Typically used to hold information needed for the current instruction, or results of the previous instruction. There are often only 8 of these, and information is immediately offloaded to either the stack or heap to make them available for the next instructions.
- Stack Pointer register: Memory address of the top of the stack.
- Flag register: Keep track of any flags, like integer overflow.

Units:

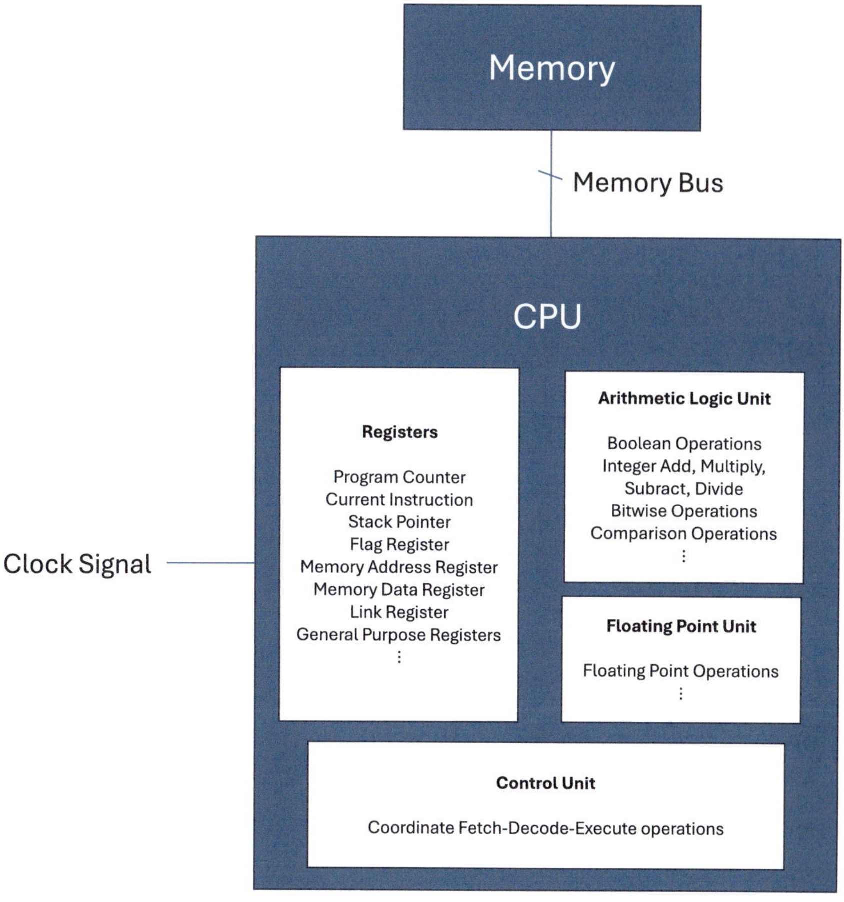

Fig. 4.15 Layout of a simple modern computer

Fig. 4.16 Fetch–Decode–Execute cycle

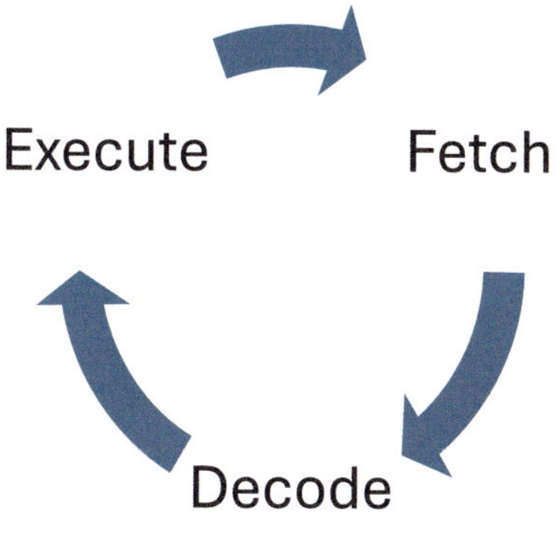

- Control Unit: Responsible for coordinating the Fetch–Decode–Execute cycle. Fetches instructions from memory, decodes them by activating the appropriate circuits, and executes the instructions (using however many clock cycles are needed for each step).
- Arithmetic Logic Unit (ALU): Circuits for binary operations (AND, OR, Bit-shift, etc.), basic arithmetic (ADD, Subtract, Multiply, Divide), etc. The ALU is coordinated by the Control Unit.
- Floating-Point Unit (FPU): Circuits for floating-point operations. It typically requires several clock cycles to complete. The FPU is also coordinated by the Control Unit.

A CPU typically has a very limited number of registers (typically 8 general-purpose registers for arithmetic, and the rest reserved for the program counter, link register, error flags, stack pointers, etc.). CPUs rely heavily on RAM to hold any data beyond the immediate operations being performed in the CPU. One may draw fuzzy parallels between the state of CPU registers with Turing machine finite state control, RAM as the tape, and the program counter as the position of the read–write head.

The rising or falling edges of a clock trigger changes in the register states of a CPU. In practice, one instruction may take several clock cycles to execute (RISC instructions require fewer cycles but have longer code; CISC instructions require more cycles to execute but have shorter code), and the fetch–decode–execute cycle is pipelined (i.e., they may be occurring simultaneously on every clock cycle!). All these operations are coordinated by the control unit in the CPU.

For brevity, we will have to gloss over additional details of how these individual units are implemented and connected, as it is complex and lengthy. By now, we hope that the reader is convinced that all these pieces can be implemented using logic gates, latches, and edge detectors, and the control unit coordinates these operations with the ticking of a clock signal.

Computer Programming

With a functioning CPU that can execute a sequence of operations (stored in RAM) that may require access to some data (also stored in RAM), one may simply write a sequence of instructions for the CPU to read information from memory, process it, and store the results in memory. We are now in the territory of computer programming.

With these basic ingredients, we can turn back to the problem of adding a vector of N integers using only one addition circuit. We use the following pseudocode to instruct a CPU to add numbers.

```
Setup:
CPU Register R3: Memory Address: First integer in vector
CPU Register R4: Memory Address: N (# of elements in vector)
```

```
Program Counter: Memory Address: Instruction 1 of Addition Program
Addition Program:
Instruction #   Instruction
Instruction 1   Load N into CPU Register R0 (using Memory Address in
R4)
Instruction 2   Write integer 0 into CPU Register R2
Instruction 3   Load number in memory address given by R3 into R1
Instruction 4   Add R2+R1, put result in R2
Instruction 5   Decrement R0 = R0-1
Instruction 6   Increment R3 = R3+1
Instruction 7   Compare R0 with 0
Instruction 8   If greater than, write address of Instruction 3
              in Program Counter Register
```

Using such a program, we can add N unsigned integers with *a single addition circuit* and a few lines of instructions that loop over the integers, as opposed to at least $N - 1$ addition circuits for a Boolean logic circuit. Note that using this same strategy, we can perform arbitrary Boolean operations by providing a sequence of instructions instead of explicitly constructing equivalent circuits.

Of course, in today's world, we typically instruct computers using programming languages that are easier to read by humans. These languages are "compiled" into machine code using compilers like gcc or "interpreted" using interpreters like Python or MATLAB.

To demonstrate the increasing level of abstraction between various programming language levels, we provide below three minimal examples of code implementing an even simpler task: adding two unsigned integers stored in memory and storing the result in memory:

- In a high-level language like Python, one may do this by writing a one-line function:

```
# Add two integers in Python
def ADDNUMS(num1, num2):
   return num1 + num2
... # Create two integers num1 and num2
sum = ADDNUMS(num1, num2)
...
```

- For a low-level language, some additional steps like declaring variables and data types are needed:

```
// Add two integers in C
#include <stdio.h>
unsigned int ADDNUMS(int num1, int num2) {
   return num1 + num2;
}
unsigned int main() {
   unsigned int num1, num2, sum;
   ... // Populate num1 and num2
   sum = ADDNUMS(num1, num2);
   ...
   return 0;
}
```

- In assembly language, one must explicitly coordinate the flow of information in the CPU registers and memory and the program counter:

```
;Add two numbers in ARM Thumb2 Assembly language
;*************************************************************
;Main Program
;*************************************************************
;LABEL     DIRECTIVE  VALUE    COMMENT
        AREA   main, READONLY, CODE
        THUMB
        EXTERN  ADDNUMS
        EXPORT  __main
__main    ...      ...     ;Start of Main program
        ...     ...     ...
        ...     ...     ;Put memory addresses in CPU registers R0: sum (not
yet computed), R1: num1, R2: num2
       BL    ADDNUMS  ;Go to instruct. number with label ADDNUMS, store
addr. of next __main instruct. in Link Register LR
       ...    ...    ;Resume Main (LR points to this instruct.). Addresses
of numbers are in R0,R1,R2.
        ...     ...     ...
        ...     ...     ...
;*************************************************************
; End of Main Program
;*************************************************************
     ALIGN
     END
;ADDNUMS code: When this code is launched, the main program is
expected to have prepared:
; 1) the memory address of num1 and num2 in CPU registers R1 and R2
```

```
; 2) the memory address of where to store sum in register R0
; 3) the memory address of the next instruction to execute after
ending program in LR (Link Register)
;This program will
; 1) Push the program state to the stack (R1, R2)
; 2) Load the two numbers into R1 and R2 using their addresses (over-
write addresses of num1, num2 with num1, num2)
; 3) Add the two numbers and store sum in register R1
; 4) Store sum (sum is in R1) in the memory address of sum (address is
in R0)
; 5) Pop R1 and R2 from stack into R1 and R2 to restore registers to
original state
; 6) Link program back to next instruction in __main
;*************************************************************
; ADDNUMS program
;*************************************************************
;LABEL    DIRECTIVE VALUE    COMMENT
          AREA      main, READONLY, CODE
          THUMB
          EXPORT    ADDNUMS
ADDNUMS   PUSH          R2      ;Push data in registers R0,R1,R2 to the
computational stack in memory
          PUSH      R1
          LDR       R1,[R1]     ;Load 32-bit integer in memory address
given by R1 into R1
          LDR       R2,[R2]     ;Load 32-bit integer in memory address
given by R2 into R2
EXAMPLE   ADD       R1,R1,R2    ;Add and store in R1: R1 = R1 + R2
          STR       R1,[R0]     ;In memory address given by R0, store
32-bit integer stored in R1
          POP       R1          ;Restore remaining registers to original
state
          POP       R2
          BX        LR          ;End of ADDNUMS, resume __main using link
register
;*************************************************************
; End of ADDNUMS program
;*************************************************************
          ALIGN
          END
```

A compiler will convert human-readable code into machine code, i.e., replace CPU instructions with the corresponding binary number for that instruction and link every LABEL in assembly code with the corresponding memory address where that

4.6.4 ADD (register)

This instruction adds a register value and an optionally-shifted register value, and writes the result to the destination register. It can optionally update the condition flags based on the result.

Encodings

T1
```
ADDS <Rd>,<Rn>,<Rm>          Outside IT block.
ADD<c> <Rd>,<Rn>,<Rm>        Inside IT block.
```

15	14	13	12	11	10	9	8 7 6	5 4 3	2 1 0
0	0	0	1	1	0	0	Rm	Rn	Rd

Fig. 4.17 ADD instruction opcode encoding, ARM Thumb-2 Supplement Reference Manual

instruction is stored. As an example, according to the *ARM Architecture Reference Manual Thumb-2 Supplement*, the instruction.

```
ADD R1,R1,R2
```

labeled as `EXAMPLE` in the assembly example above will be converted to a 16-bit machine opcode:

```
0001100010001001
```

by a compiler according to the datasheet excerpt shown in Fig. 4.17. This string of zeros and ones will activate the appropriate registers and ALU circuits in the CPU and activate the electrical connect between the registers and the ALU. The available set of instructions, registers, and their corresponding opcodes are provided by the manufacturer of any processor in developer manuals.

One rarely writes code in assembly or machine opcodes due to its tremendous difficulty, relying on compilers to deal with at least low-level code, and, more often these days, high-level code. However, knowledge of the inner workings of digital computers allows one to better appreciate their power, limitations, and potential.

Progress in Classical Computing

This discussion barely scratches the surface of modern-day CPUs, which require several other layers of abstraction, e.g., CPU cache. The aim of this discussion is to provide the reader with an abstract understanding of how classical computing is done electronically using the binary number system from the program level down to electronic circuits executing instructions at the transistor level. Modern-day computing is highly complex, utilizing several advanced features like multiple levels of cache, operating systems, dynamic memory allocation, multi-threading, pipelining, multiple cores, speculative execution, RAM, disk storage,

GPU accelerators, FPGAs, network connections, and many other capabilities. However, at its heart lie basic components like logic gates, clock signals, and memory cells. Although modern electronic computational devices differ from true Turing machines, one may prove that a digital electronic computer with (hypothetical) infinite memory is equivalent to a Turing machine. It is fortunate that for many problems of practical interest finite memory is sufficient.

For readers interested in learning more about practical modern computing, one may start with studying digital design [2] and then proceeding to read about computer architectures [3].

We end our discussion with how progress in computing has scaled over the past decades, and how it is expected to proceed in the future: Moore's law, and the GPU version of Moore's law.

Over the years the transistor count in CPUs has roughly doubled every two years, and this principle is referred to as *Moore's Law* [4] as illustrated in Fig. 4.18. This doubling of transistors has roughly translated to a doubling of the computational capabilities of processors. This is attributed to steady advancements in transistor and manufacturing technologies, enabling smaller "feature sizes" of integrated circuits. However, this approach of miniaturization is approaching its limits; feature sizes smaller than 1 nm are prone to quantum effects, making the devices lose their deterministic behavior (at the time of writing this book, 2nm features enabled by extreme ultraviolet lithography are state-of-the-art [5]).

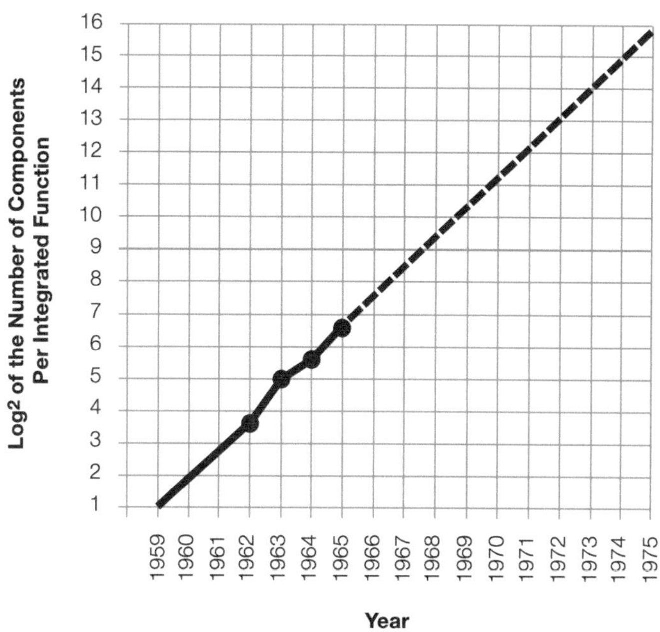

Fig. 4.18 Gordon Moore's prediction in his original publication [4]

Beyond the issues of feature sizes, the operable clock frequencies of CPUs have stopped increasing due to heat dissipation issues, which can lead to a *compute bound* situation. Moreover, programs are often *memory bound*, i.e., the limited bandwidth between the CPU and memory inhibits speedups since the CPU idles while data is being read from (much slower) memory. In the realms of distributed memory, heterogeneous computing, and parallel processing data may need to be moved between memories or messages/signals may need to be sent between computing units. It is not rare for such computing systems to become *communication bound*.

Several of these concerns have been alleviated by parallel computing enabled by modern GPUs. Parallel programming enables speedups by performing many operations in parallel, enabling speedups for many practical problems. Revisiting our problem of adding N integers, a CPU will require $O(N)$ CPU clock cycles, while a GPU with $N - 1$ processors can perform the same task in $O(\log N)$ GPU cycles using parallel reduction in software. The explosive growth of GPU computing has posited *Huang's law* [6]: the number of transistors in GPU devices more than doubles every year. A single modern GPU may have thousands of computational cores; multiple GPUs can process information on a single computer; multiple computers can be connected to form clusters or supercomputers.

However, many computational tasks in scientific computing cannot benefit from parallelization. A very basic example, Newton iterations to compute the roots of a function $f : \mathbb{R} \to \mathbb{R}$:

$$x_{n+1} = x_n + \frac{f(x_n)}{f'(x_n)}$$

require sequential computation of the iterates and will be limited by clock speeds.

Even for problems that are not completely sequential, parallelization is far from a panacea. Problems which are embarrassingly parallel, i.e., require no communication between processes, are well-suited for parallelization over multiple processors. For all other problems, the speedup possible from parallelization will be limited by the communication overhead between processors, which may be on the same CPU or GPU, or may span multiple CPUs, GPUs, computer nodes, or even be distributed across the Internet.

Beyond the ability to compute fast, one may also be restricted by the amount of available memory and energy consumption. Simulating quantum mechanics is exponentially expensive in general and is simply intractable on classical computers. The quantum equivalent of Moore's law has not been established yet, albeit some suggestions have been floated, e.g., quantum volume [7–9].

References

1. M.A. Nielsen, I.L. Chuang, *Quantum Computation and Quantum Information: 10th Anniversary Edition*, 1st ed. (Cambridge University Press, 2012). https://doi.org/10.1017/CBO9780511976667

2. M.M. Mano, *Digital design*, 2nd edn. (Prentice-Hall International, London, 1991)
3. J.L. Hennessy, D.A. Patterson, *Computer architecture: a quantitative approach*, Fifth ed. (Amsterdam Heidelberg, Elsevier, Morgan Kaufmann, 2012)
4. G.E. Moore, Cramming more components onto integrated circuits, Reprinted from Electronics, volume 38, number 8, April 19, 1965, pp.114 ff. IEEE Solid-State Circuits Soc. Newsl. **11**(3), 33–35 (2006). https://doi.org/10.1109/N-SSC.2006.4785860
5. D. Kazazis, J.G. Santaclara, J. Van Schoot, I. Mochi, Y. Ekinci, Extreme ultraviolet lithography. Nat Rev Methods Prim. **4**(1), 84 (2024). https://doi.org/10.1038/s43586-024-00361-z
6. T.S. Perry, Move over, Moore's law. Make way for Huang's law [Spectral Lines]. IEEE Spectr. **55**(5), 7–7 (2018). https://doi.org/10.1109/MSPEC.2018.8352557
7. C.H. Baldwin, K. Mayer, N.C. Brown, C. Ryan-Anderson, D. Hayes, Re-examining the quantum volume test: Ideal distributions, compiler optimizations, confidence intervals, and scalable resource estimations. Quantum **6**, 707 (2022). https://doi.org/10.22331/q-2022-05-09-707
8. K. Miller, C. Broomfield, A. Cox, J. Kinast, B. Rodenburg, An improved volumetric metric for quantum computers via more representative quantum circuit shapes (2022). arXiv:2207.02315. https://doi.org/10.48550/arXiv.2207.02315
9. N. Moll et al. Quantum optimization using variational algorithms on near-term quantum devices. Quantum Sci. Technol. **3**(3), 030503 (2018). https://doi.org/10.1088/2058-9565/aab822

5 Information and Complexity Theory

In the previous two chapters, we have learned about the theoretical foundations of computing and how practical classical computing machines operate. Using the concept of a deterministic Turing machine, we have explored the idea of what is computable and have concluded that any problem with a solution described as an algorithm is theoretically solvable, and that deterministic Turing machines can execute any algorithm. We noted that Turing machines with multiple tapes and non-deterministic Turing machines can theoretically solve the same problems, albeit in a more efficient manner. We then provided an overview of how classical computers operate, recognizing that memory is finite and limited, and that the number of computational steps that can be performed within a finite time is bounded by practical constraints, such as clock speeds, parallelizability, communication overheads, and memory access speeds.

We now turn our attention to study the tractability of problems under the lens of their complexity, i.e., the time and resources needed to execute algorithms, and classify them. This enables us to allocate time and computational resources to tractable problems and direct computational resources toward alternatives for intractable problems, seeking approximate solutions.

Classical Decision Problems and Complexity Classes

We touch upon the idea of a *decision problem* in this chapter, i.e., using designated accept or reject states, a deterministic (or a non-deterministic) Turing machine can *decide* a problem as accept/yes/affirmative or reject/no/negative. Problems are solved using an algorithm, which can be efficient (polynomial resources) or inefficient (superpolynomial resources).

O. M. Raisuddin and S. De, *Quantum Computing for Engineers*,
https://doi.org/10.1007/978-3-032-03325-3_5

The *time* (number of steps) and *space* (memory) required to *decide* problems is one important factor in classifying them. Another important factor is the notion of *verifiers* and *certificates*. To understand the relations between various problems, we will also introduce the concept of *reducing* a problem to another. In this chapter, we will use the traveling salesman problem to indicate the differences between problems and their complexities.

Let's first consider the problem of finding the maximum number in a set. The *optimization* version of this problem can be stated as

Given a set of integer numbers $A \subset \mathbb{Z}$, find $y \in A$ s.t. $y \geq z \; \forall \; z \in A$.

We can state the *decision* version of this problem as follows:

Given a set of integer numbers $A \subset \mathbb{Z}$ and $z \in \mathbb{Z}$, does there exist a $y \in A$ s.t. $y \geq z$?

It is straightforward to see that finding the maximum (optimization version) in a set of numbers will take $O(n)$ steps. For the decision version, we may simply scan through the set to seek $y \geq z$, requiring $O(n)$ steps at most. For the decision version of the problem, the *accept* case will be accompanied by a *certificate*, which can be some y satisfying $y \geq z$. The *verifier* is an algorithm that will verify this certificate. For both versions of this problem, the verifier can go through the set to find either the maximum or find $y \geq z$, which is pretty much the decision algorithm itself! Both the algorithm to decide the decision problem and verify the certificate require a polynomial (of degree 1) number of steps. Therefore, the decision problem is decidable and verifiable in polynomial time and belongs to the complexity class P. We can also solve the optimization version of this problem in polynomial time, but the class P is only defined for decision problems.

Note that if A is a tuple of integers, the certificate could instead be the index of y in the tuple A. In this setting the verification can be done in constant $O(1)$ time for the decision version of the problem, while the optimization problem still requires $O(n)$ time.

Now let's investigate traveling salesman problems.

One version of this problem is a *decision* problem:

Given a list of N cities and the distances between them, does there exist a path such that each city is visited exactly once with the same starting and final city, and the total distance is at most k?

A deterministic or non-deterministic Turing machine can *decide* this problem by answering in the *affirmative* or *negative*. The best-known algorithms to decide this problem with a deterministic Turing machine require exponential time in the worst case. However, a non-deterministic Turing machine can decide this problem in polynomial time.

In either case, we will also be provided a *certificate* for the *affirmative* case in the form of the path that is shorter than k. Using a *verifier*, we can use the *certificate* to confirm that the solution is indeed affirmative. The length of the path (a sum of N distances) can be verified in $O(N)$, polynomial time, using a deterministic Turing machine.

Since the certificate is verifiable in polynomial time by a deterministic Turing machine, this decision problem belongs to the *non-deterministic polynomial time* class NP. By this definition, the class of problems P also lies in NP. An alternative and equivalent definition for NP is the class of problems decidable by non-deterministic Turing machines in polynomial time. It is widely accepted, but not proven that $P \subset NP$ and $P \neq NP$.

Now we turn our attention to a special subset of problems in NP: NP-Complete problems. Every problem in NP can be *reduced* to a problem in the class of NP-Complete problems, where the reduction is efficient, requiring at most polynomial time. By reducing any problem in NP to any NP-complete problem, one may solve the equivalent NP-complete problem to implicitly solve the NP problem. These decision problems are therefore the "hardest" problems in the class NP; solving these efficiently (in polynomial time using a Turing machine) solves all problems in NP efficiently.

Another version of the traveling salesman problem is an *optimization* problem:

Given a list of N cities and the distances between them, what is the shortest path such that each city is visited exactly once, with the same starting and final city?

This is not a decision problem; it simply asks for an optimal solution. Even if any Turing machine provides an optimal solution, we do not have any method to verify that it is the optimum solution other than seeking the optimal solution itself, which requires exponential time.

We will now finally introduce the NP-hard class of problems. The class NP-hard is not restricted to decision problems. These problems are at least as hard as NP-complete problems (the hardest problems in NP).

The conjectured relation among the decision problems NP and the general class of problems NP-hard is shown in Fig.5.1.

Until now, we have mostly considered problems that are decidable by non-deterministic Turing machines within polynomial time, and by deterministic Turing machines within exponential time. The space required for these problems is polynomial since a deterministic Turing machine can simply try out every possible solution one by one exhaustively, i.e., they belong to the class PSPACE. A broader class of problems is problems that require exponential time, even for

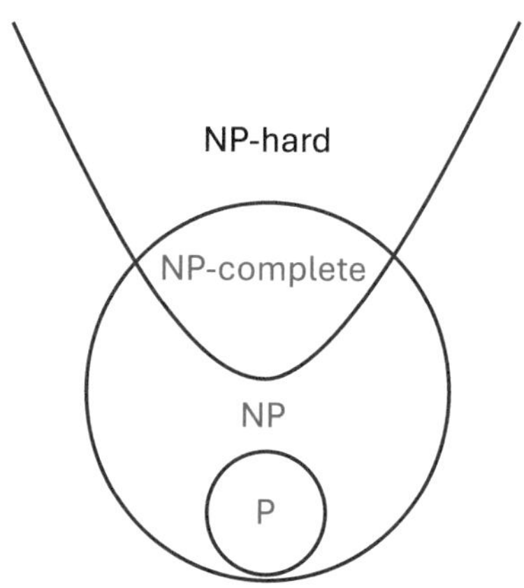

Fig. 5.1 Conjectured relation among P, NP, NP-complete, and NP-hard classes of problems

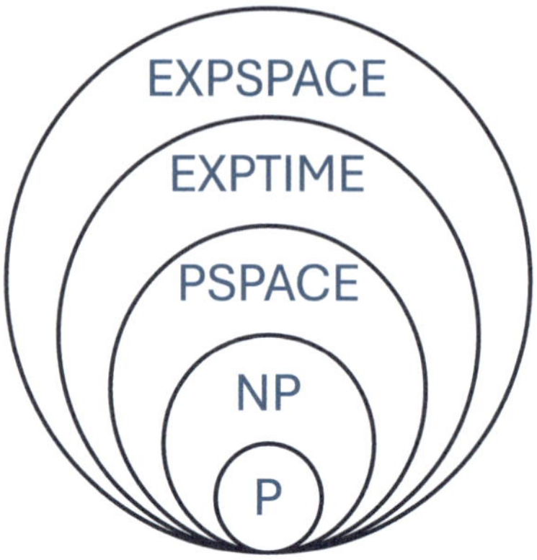

Fig. 5.2 Conjectured relation among broader classes of decision problems

a non-deterministic Turing machine, and problems that require an exponential amount of memory, which are the classes EXPTIME and EXPSPACE, respectively. These decision problems are nested as P $\subseteq$ NP $\subseteq$ PSPACE $\subseteq$ EXPTIME $\subseteq$ EXPSPACE, as visualized in Fig. 5.2.

Probabilistic and Quantum Complexity Classes

We will now define a few more complexity classes and will establish the conjectured relation between the power of classical computers and quantum computers.

Bounded-Error Probabilistic Polynomial (BPP): Problems that can be decided by a probabilistic Turing machine in polynomial time with a probability of correctly accepting a problem $\geq \frac{2}{3}$ and a probability of incorrectly accepting a problem $\leq \frac{1}{3}$ for all instances.

Probabilistic Polynomial (PP): Problems that can be decided by a probabilistic Turing machine in polynomial time with a probability of correctly accepting a problem $> \frac{1}{2}$ and a probability of incorrectly accepting a problem $\leq \frac{1}{2}$ for all instances.

The difference between PP and BPP lies in the fact that for BPP, the algorithm can simply be repeated a few times to (exponentially) increase the probability of correctly deciding the input. For PP, this guarantee is lost. Naturally, BPP $\subseteq$ PP.

Quantum computers are probabilistic devices. We have not yet introduced quantum computing, but at this point it is sufficient to know that a quantum computer executes a *quantum circuit* prepared by a classical computer. The size of this quantum circuit determines the time required for its execution. Quantum circuits, which are considered feasible, are polynomial in size. A quantum computer can probabilistically decide on an input. We define the class of problems BQP now as follows.

Bounded-Error Quantum Polynomial (BQP): Problems that can be decided by a quantum computer in polynomial time with a probability of correctly accepting a problem $\geq \frac{2}{3}$ and a probability of incorrectly accepting a problem $\leq \frac{1}{3}$ for all instances.

The circuit that the quantum computer runs must be generable (and executable) within polynomial time.

We will now summarize our entire discussion. For classical computing, problem complexities are nested as.

P $\subseteq$ NP $\subseteq$ PP $\subseteq$ PSPACE $\subseteq$ EXPTIME $\subseteq$ EXPSPACE

For classical probabilistic and quantum computing problems, complexities are nested as

P $\subseteq$ BPP $\subseteq$ BQP $\subseteq$ PP $\subseteq$ PSPACE $\subseteq$ EXPTIME $\subseteq$ EXPSPACE

which is summarized in Fig. 5.3.

From these classes, we can get a better understanding of quantum computers and their power and efficiency. Since BQP $\subseteq$ EXPTIME, quantum computers are at most exponentially more efficient than classical computers, and since BPP $\subseteq$ BQP computers are at least as powerful as classical computers. What is surprising

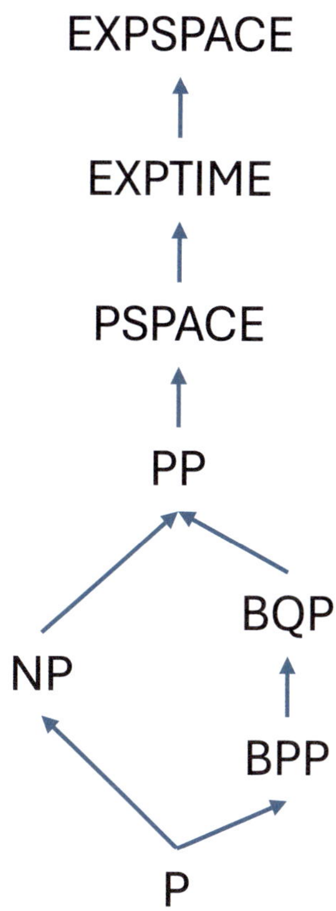

Fig. 5.3 Conjectured nesting relation among NP, BPP, BQP, and PP. The arrows indicate subsets, i.e., BQP $\subset$ PP

is that for decision problems, BQP $\subseteq$ PSPACE, i.e., a classical computer can, given enough time, decide any problem decidable by a quantum computer using polynomial space. For more detailed analyses with more classes, e.g., MA and QMA, we refer the readers to [1, 2].

Information is Physical

The concepts of information and computing are implicitly connected to physics. We start our discussion with a thought experiment leading to a paradox: Maxwell's Demon [3].

Consider a box with particles bouncing around, similar to a gaseous phase of matter. The box is divided into two chambers FAST and SLOW, with a door operated by a "demon." The demon keeps track of all the particles by measuring their velocities, and opens and shuts a hypothetical door in such a manner that fast-moving particles are allowed to move from chamber SLOW to FAST, and slow-moving particles are allowed to move from chamber FAST to SLOW. The door remains shut otherwise.

By doing this long enough, most, or all, of the particles will eventually be separated into FAST and SLOW, i.e., hot and cold gases, respectively, and the entropy of the system will have decreased as illustrated in Fig. 5.4. Considering that the door is small enough and the energy expended in tracking the particles and operating the door is negligible, one may then (fallaciously) consider operating a heat pump to obtain "free" energy in violation of the second law of thermodynamics.

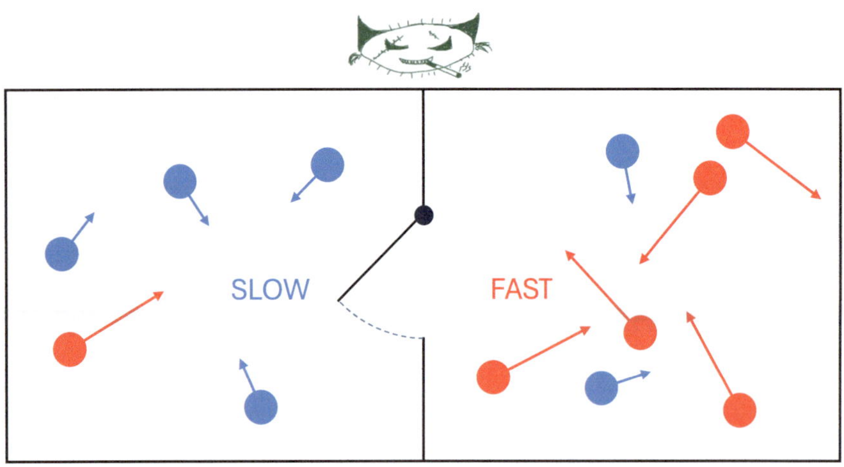

Fig. 5.4 Maxwell's "demon" separating slow (blue) and fast (red) particles into FAST and SLOW chambers

We know that violating the second law of thermodynamics is impossible: the total entropy of a closed system cannot decrease. Although there are many troublesome idealizations in this thought experiment, one scenario is that even if the demon can measure the velocities of the particles and operate an ideal door, the velocity information will have to be tracked and stored. Eventually, old information will have to be deleted (either during one separation cycle or any subsequent separation cycles). This act of erasing previously known information can be directly considered an increase in entropy. Any decrease in entropy as a result of separating the particles will be offset by the erasure of information.

Landauer's principle formalizes this concept. For interested readers, we note that this is a nuanced topic and refer them to [4] for an excellent exposition on this subject. With a preamble, we have set the stage for the remainder of this book.

References

1. R. Manenti, M. Motta, *Quantum Information Science* (Oxford University Press, Incorporated, Oxford, 2023)
2. J. Watrous, Quantum Computational Complexity, in *Encyclopedia of Complexity and Systems Science*, ed. by R.A. Meyers (Springer New York, New York NY, 2009), pp. 7174–7201. https://doi.org/10.1007/978-0-387-30440-3_428
3. The Sorting Demon of Maxwell 1. Nature **20**(501), 126–126 (1879). https://doi.org/10.1038/020126a0
4. C.H. Bennett, Notes on Landauer's principle, reversible computation, and Maxwell's Demon. Stud. Hist. Philos. Sci. Part B: Stud. Hist. Philos. Mod. Phys. **34**(3), 501–510 (2003). https://doi.org/10.1016/S1355-2198(03)00039-X

Part II
A Brief Introduction to Quantum Mechanics

This part provides a concise introduction to the essential concepts and experiments underlying quantum mechanics, presented specifically for computational engineers and applied scientists. It is not meant to serve as a comprehensive quantum mechanics course, but rather as a focused overview that highlights key physical phenomena and foundational principles relevant to quantum computing.

We emphasize physical intuition and conceptual clarity over exhaustive theoretical treatments. Readers seeking deeper exploration or rigorous derivations are encouraged to consult standard quantum mechanics textbooks. The topics introduced here set the stage for understanding quantum computing concepts presented in subsequent parts.

Chapter 6, "A Gentle Introduction to Quantum Mechanics", introduces the structure of quantum theory through foundational concepts, including quantum states, superposition, measurement, and probability amplitudes. Emphasis is placed on developing intuition for how quantum systems behave differently from classical systems, with a minimal use of mathematical formalism.

Chapter 7, "The Stern–Gerlach Experiment", presents the Stern–Gerlach experiment as a concrete illustration of measurement, state collapse, and spin quantization. The chapter illustrates how discrete measurement outcomes emerge in quantum systems and how they are related to the mathematical structure of quantum states.

Chapter 8, "Photon Polarization", uses polarization states of photons to reinforce and generalize earlier concepts. The chapter explores basis changes and probabilistic outcomes in the context of light, providing an accessible experimental analogy for qubit operations.

A Gentle Introduction to Quantum Mechanics

6

The theories of physics that were discovered and studied up to the end of the nineteenth century are what we now call classical physics. These include and are not limited to Newton's laws, continuum mechanics, and Maxwell's equations for electromagnetism. These frameworks describe a wide range of macroscopic phenomena with remarkable accuracy. However, as experimental techniques advanced, discrepancies began to emerge between classical predictions and empirical observations, particularly at atomic and subatomic scales. These failures ushered in the era of modern physics, characterized by two revolutionary frameworks: relativity and quantum mechanics.

One of the crises that was resolved by quantum mechanics is the "ultraviolet catastrophe." According to the classical Rayleigh–Jeans law describing the frequency spectrum of a black body at a given temperature, a black body would emit unbounded energy in the ultraviolet spectrum and higher frequencies:

$$B_\nu(T) = \frac{2\nu^2 k_B T}{c^2}$$

where $B_\nu(T)$ is the intensity of the frequency ν for a blackbody at equilibrium temperature T and k_B, c are the Boltzmann constant and the speed of light respectively. As $\nu \to \infty$, $B_\nu \to \infty$, which is unphysical and disagrees with experiments.

This paradox was resolved by Max Planck, who proposed that electromagnetic energy is emitted in discrete packets, or quanta [1]. These ideas were later extended by Albert Einstein to explain the photoelectric effect and by Niels Bohr to develop models for the atom. These developments culminated in the formulation of quantum mechanics, a new theoretical framework capable of accurately describing atomic and subatomic phenomena.

O. M. Raisuddin and S. De, *Quantum Computing for Engineers*,
https://doi.org/10.1007/978-3-032-03325-3_6

Quantum mechanics is governed by a set of mathematical postulates, analogous to the foundational laws of classical mechanics. These postulates define how quantum systems are represented, how observables are associated with operators, and how measurements and time evolution are described.

Postulate 1—Wavefunctions:

The state of a quantum mechanical system is completely specified by a wavefunction $|\Psi\rangle$ *in a complex Hilbert space such that* $\langle\Psi \mid \Psi\rangle = 1$.

Postulate 2—Observables:

For every observable property of a quantum mechanical system described by $|\Psi\rangle$, *there exists a linear Hermitian operator A. The operator encodes all possible measurement outcomes for that observable.*

Postulate 3—Measurements:

The outcomes of any measurement of an observable A will be limited to the eigenvalues λ_a *of A.*

Postulate 4—Expectation Values:

The expected value of an observable A for a system in state $|\Psi\rangle$ *is given by* $\langle\Psi|A|\Psi\rangle$.

The probability of obtaining an outcome λ_a *with a corresponding eigenvector* $|a\rangle$ *is*

$$p(a) = \langle\psi|(|a\rangle\langle a|)|\psi\rangle$$

Postulate 5—Time Evolution:

The wavefunction of a quantum system evolves in time according to the time-dependent Schrodinger equation:

$$ih\frac{d}{dt}|\Psi(t)\rangle = H|\Psi(t)\rangle$$

where h is Planck's constant and H is the Hamiltonian for the system.

Postulate 6—Wavefunction Collapse:

Upon measuring an observable A and obtaining an eigenvalue λ_a, *the system's state collapses to the corresponding eigenvector* $|a\rangle$.

Although quantum mechanics has profound physical implications, its mathematical formalism is relatively straightforward, relying primarily on linear algebra and probability theory. The key challenge is not mastering the mathematics but developing intuition for how quantum systems behave—how they evolve, interfere, and respond to measurement.

To build that intuition, the next two chapters present two simple but foundational experiments—the Stern–Gerlach experiment and photon polarization—framed for accessibility and insight rather than rigor. Readers are encouraged to read this part with a focus on conceptual understanding, not on fully internalizing every equation. For a deeper and more formal treatment, standard textbooks in quantum mechanics are recommended [2].

References

1. M. Planck, Kritik zweier Sätze des Hrn. W. Wien. Annalen der Physik **308**(12), 764–766 (1900). https://doi.org/10.1002/andp.19003081215
2. J.S. Townsend, *A modern approach to quantum mechanics*, 2nd ed. (University Science Books, Mill Valley, Calif, 2012)

7 The Stern–Gerlach Experiment

The Stern–Gerlach experiment [1] and its modifications help describe some basic concepts of quantum mechanics. We start our discussion by introducing a single Stern–Gerlach device.

Beam Source

The input to a Stern–Gerlach device is a beam of silver atoms. This is achieved by evaporating silver using an electric furnace in a vacuum and allowing the gas to escape through a small aperture, as shown in Fig. 7.1.

The reason for using silver is that it has one unpaired valence electron. We are looking at this experiment in retrospect through the lens of modern quantum theory; the original Stern–Gerlach experiment was designed to prove the "old quantum theory" [2] which has not withstood experimental scrutiny.

Electron Spin

Electrons carry an intrinsic property of "spin." The term spin is a misnomer; it is quite unlike the classical spin of objects with mass [3]. The reason this intrinsic property is called spin is that it exhibits behavior similar to an object having angular momentum.

Depending on the direction of their spin, electrons will deflect in a magnetic field. All the electrons in silver atoms, except the unpaired 5s1 electron, are paired as shown in Fig. 7.2 according to the *aufbau* rule. Paired electrons cancel out the

O. M. Raisuddin and S. De, *Quantum Computing for Engineers*,
https://doi.org/10.1007/978-3-032-03325-3_7

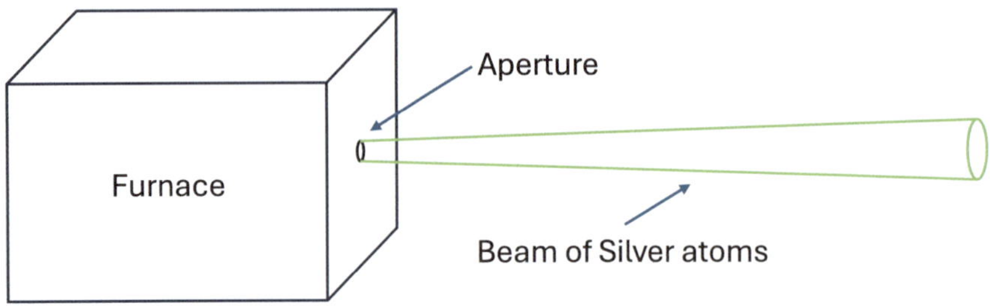

Fig. 7.1 Beam source for a Stern–Gerlach Experiment. A furnace heats silver particles, which are emitted through a small aperture as a beam of particles

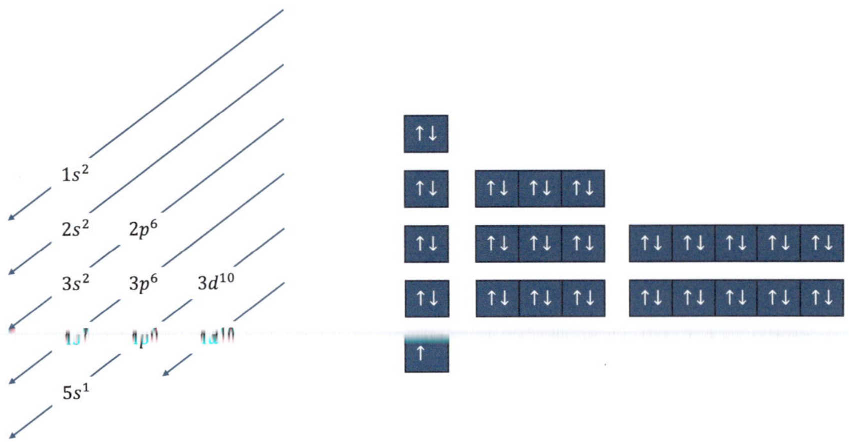

Fig. 7.2 Pairing of electrons in silver atoms in shells according to the aufbau principle

net effect of particles with opposite spins in a magnetic field. The unpaired electron, however, should cause a silver atom to deflect in the presence of a magnetic field depending on which direction the spin is aligned.

Stern–Gerlach Device and Detector

Inside the Stern–Gerlach device is a magnetic field created using magnetic poles with a special shape: one pointed pole and one flat pole as shown in Fig. 7.3. The magnetic field created by this arrangement is inhomogeneous, or spatially varying, in the $x - z$ plane. This inhomogeneous field should deflect the silver atoms according to the angle formed between the magnetic field lines and the magnetic dipole of the unpaired electron. In this case, it deflects the electrons toward either magnet according to the alignment of the electron's spin.

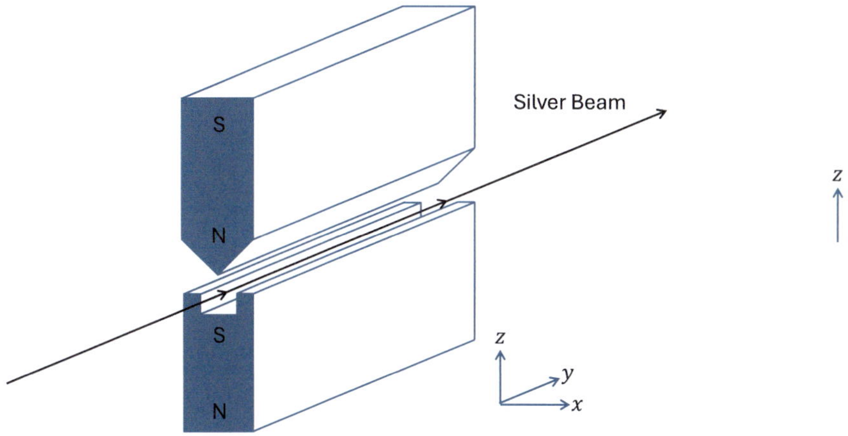

Fig. 7.3 Left: Silver beam passing through a Stern–Gerlach device. Right: Schematic of the inhomogeneous magnetic field inside a Stern–Gerlach device

After passing through the magnetic field, the silver atoms collide with a "detector," which in the original experiment was simply a metallic plate on which the silver atoms were deposited and could be observed at the end of the experiment.

Stern–Gerlach Experiments

Experiment 1

Since the particles coming from the furnace are not arranged in any specific order or aligned in any direction, one may expect that the orientation of the spin for the unpaired electron should be random. Based on this intuition, a continuum of deflections for the particles is to be expected. The surprising result is that the beam of silver atoms is split into two, as shown in Fig. 7.4. This experiment hints that the spin property of electrons is *quantized*. However, this phenomenon of quantization alone does not entirely capture the counterintuitive nature of quantum mechanics.

We now label this arrangement of magnets with the magnetic field lines parallel to the Z-axis, a Stern–Gerlach-Z (SGz) device. By rotating the device 90° around the Y-axis, we can create a Stern–Gerlach-X (SGx) device. This experiment is restated in a schematic form in Fig. 7.5 where F denotes the intensity of the beam entering the device. The intensity of the beam split in two is denoted by $F/2$ as it is halved.

We use this schematic to describe further experiments that demonstrate the counterintuitive nature of quantum mechanics.

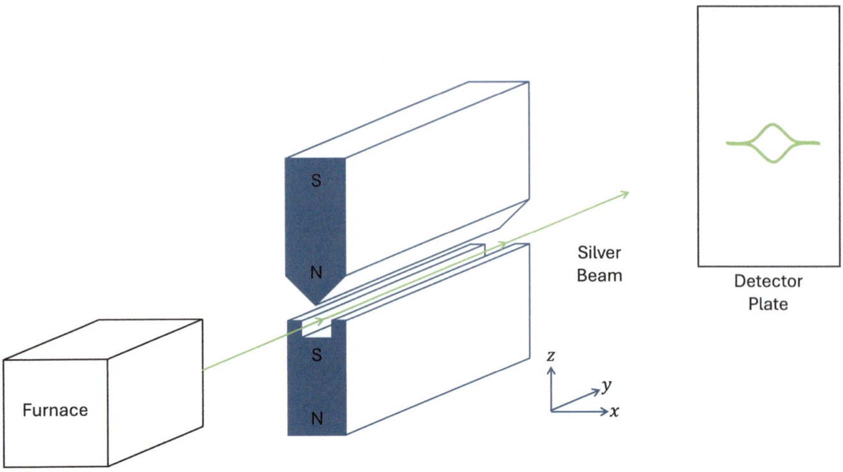

Fig. 7.4 The Stern–Gerlach device splits a beam into two distinct beams

Fig. 7.5 Schematic of a Stern–Gerlach device splitting a beam

Experiment 2

In our second experiment, shown schematically in Fig. 7.6, we feed the beam deflected in the +Z direction by an SGz device into another SGz device. Since the particles exiting each of the beams of the SGz device are aligned in the +Z direction, we expect the second SGz device to once again deflect those particles in the +Z. Indeed, this is what is observed experimentally.

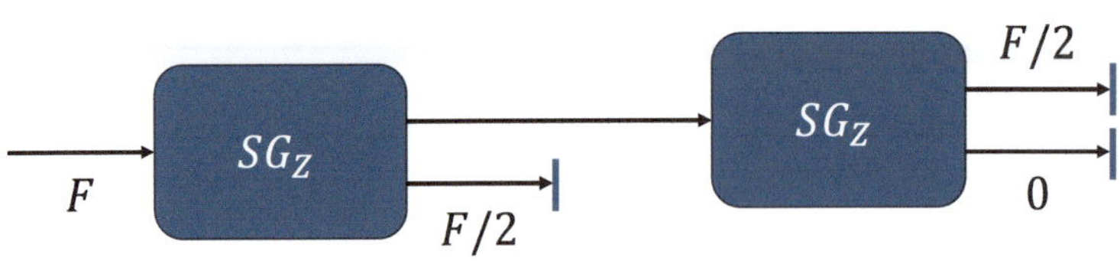

Fig. 7.6 Two Stern–Gerlach devices aligned with the same (Z) axis

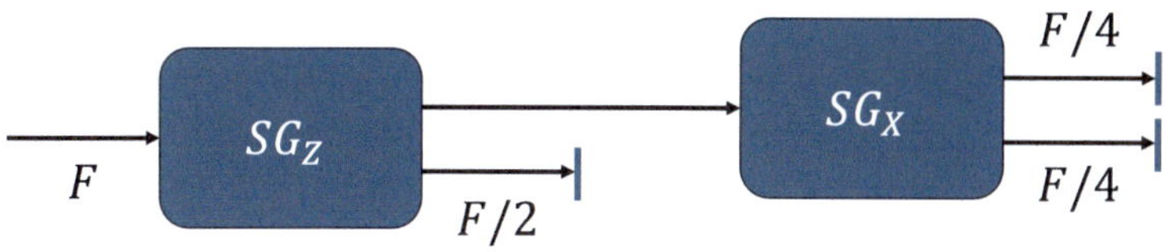

Fig. 7.7 Two Stern–Gerlach devices are aligned with mutually orthogonal axes

Experiment 3

We now replace the second SGz device with an SGx device, as shown in Fig. 7.7. The SGx device is identical to an SGz device, except that it is rotated 90° in the x–z plane. It is observed that the SGx device further splits the beam into two, based on the X component of spin.

This is not too surprising, as one may make an analogy to a spinning top forming an angle of 45° with the Z and X axes; the beam deflected in the +X direction by the SGx device had a quantized spin component of both +Z and +X.

To model each silver particle passing through these devices, one could perhaps assign a probability of $\frac{1}{2}$ to the +Z and +X property of spin for particles in the beam exiting the furnace. Furthermore, the SGz and SGx devices can be hypothesized to be "filtering" the beam into beams with +Z and +X properties, respectively. To test this hypothesis, we proceed with the next experiment.

Experiment 4

Based on the hypothesis from the previous experiment, one would expect that passing a beam deflected (filtered) in the +Z and +X directions once more through an SGz device would deflect the beam in the +Z direction once more. This is not observed experimentally. As shown in Fig. 7.8, the beam is *once again* split in the +Z and -Z directions!

These experiments suggest that the properties that deflect the beam in the x and z directions are not independent (or not orthogonal in a mathematical sense). These results necessitate the use of *probability amplitudes* to model the state of particles in the beam, which take on values within the unit disc in the complex plane, instead of probabilities, which are real numbers between 0 and 1.

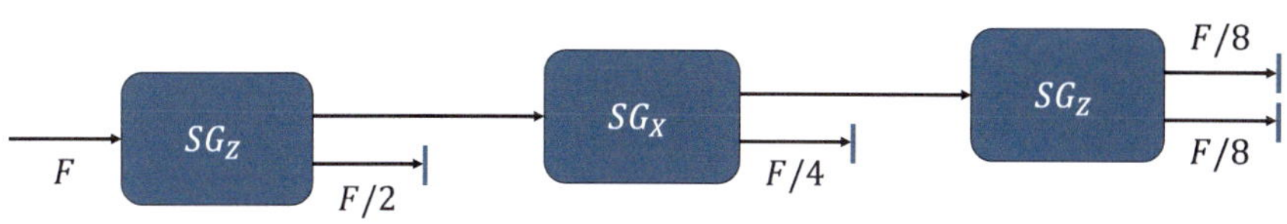

Fig. 7.8 Three Stern–Gerlach devices are aligned with alternating orthogonal axes

We now restate these results mathematically. In our statement, we do not provide a rigorous derivation of a quantum mechanical framework from experimental results; such derivations can be found in numerous quantum mechanical textbooks. Providing such a detailed discussion is not only lengthy and redundant but may also be of little use or interest to the target audience of this book. Instead, we provide a mathematical description that "works" to foster an intuition of quantum mechanics.

We first define an *orthogonal* basis $\{|+z\rangle, |-z\rangle\}$ for the property of spin based on deflection in the +Z and -Z directions. Note that this differs significantly from the vectors $+z, -z + z, -z + z, -z$ in Euclidean space which are *not orthogonal*. Bra-ket notation allows one to avoid this confusion by describing quantum mechanical objects in a separate notation.

We may model the particles exiting the electric furnace and entering the SGz device in Experiment 1 as the quantum state

$$|\psi\rangle = \frac{1}{\sqrt{2}}|+z\rangle + \frac{1}{\sqrt{2}}|-z\rangle$$

where the phase is omitted for simplicity. After exiting the device, we observe that particles are deflected in the +Z and -Z directions with probabilities $\frac{1}{2}$ each. From this we deduce that

$$p_1(|+z\rangle) = |\langle +z \mid a\rangle|^2 = \frac{1}{2}$$

$$p_1(|-z\rangle) = |\langle -z \mid a\rangle|^2 = \frac{1}{2}$$

The splitting action of the SGz device may be modeled as the projectors $|+z\rangle\langle +z|$ and $|-z\rangle\langle -z|$ for the +Z and −Z deflections, respectively.

Now we can model the states of particles in the two beams exiting the SGz device:

+Z Beam: $(|+z\rangle\langle +z|)|\psi = \frac{1}{\sqrt{2}}|+z\rangle$.

-Z Beam: $(|-z\rangle\langle -z|)|\psi = \frac{1}{\sqrt{2}}|-z\rangle$.

According to Experiment 2, if we now send a particle in the beam deflected in the +Z direction into another SGz device and measure the intensity of the beam we will get

$$p_2(|+z\rangle, |+z\rangle) = \frac{1}{2}$$

$$p_2(|+z\rangle, |-z\rangle) = 0$$

This is apparent by computing

+Z, +Z Beam: $(|+z\rangle\langle +z|)(|+z\rangle\langle +z|)|\psi\rangle = \frac{1}{\sqrt{2}}|+z\rangle$.

+Z, -Z Beam: $(|-z\rangle\langle -z|)(|+z\rangle\langle +z|)|\psi\rangle = 0$.

To model Experiment 3, we first need to introduce an orthogonal basis for the property of spin based on deflection in the +X and -X directions:

$$\{|+x\rangle, |-x\rangle\}$$

where $|+x\rangle = \frac{1}{\sqrt{2}}|+z\rangle + \frac{1}{\sqrt{2}}|-z\rangle$ and $|-x\rangle = \frac{1}{\sqrt{2}}|+z\rangle - \frac{1}{\sqrt{2}}|-z\rangle$.

Similar to the $\{|+z\rangle, |-z\rangle\}\{|+z\rangle, |-z\rangle\}$ basis, the effect of the SGx device may be modeled as the projectors $|+x\rangle\langle+x|$ and $|-x\rangle\langle-x|$ for the +X and -X deflections respectively.

According to experimental results, for the beam deflected in the +Z direction and further split by the SGx device:

$$p_3(|+z\rangle, |+x\rangle) = \frac{1}{4}$$

$$p_3(|+z\rangle, |-x\rangle) = \frac{1}{4}$$

These probabilities are apparent by computing the corresponding probability amplitudes:

+Z, +X Beam: $(|+x\rangle\langle+x|)(|+z\rangle\langle+z|)|\psi\rangle = \frac{1}{2}|+x\rangle$.

+Z, -X Beam: $(|-x\rangle\langle-x|)(|+z\rangle\langle+z|)|\psi\rangle = \frac{1}{2}|-x\rangle$.

Finally, we can model Experiment 4 similarly as

+Z, +X, +Z Beam: $(|+z\rangle\langle+z|)(|+x\rangle\langle+x|)(|+z\rangle\langle+z|)|\psi\rangle = \frac{1}{2\sqrt{2}}|+x\rangle$.

+Z, +X, -Z Beam: $(|-z\rangle\langle-z|)(|+x\rangle\langle+x|)(|+z\rangle\langle+z|)|\psi\rangle = \frac{1}{2\sqrt{2}}|-x\rangle$.

Corresponding to the experimentally observed probabilities

$$p_4(|+z\rangle, |+x\rangle, |+z\rangle) = \frac{1}{8}$$

$$p_4(|+z\rangle, |+x\rangle, |-z\rangle) = \frac{1}{8}$$

As we can see, quantum mechanics successfully models these non-intuitive experimental observations.

References

1. W. Gerlach, O. Stern, Der experimentelle Nachweis der Richtungsquantelung im Magnetfeld. Z. Physik, **9**(1), pp. 349–352, (1922), https://doi.org/10.1007/BF01326983.
2. H. Johnston, How the Stern–Gerlach experiment made physicists believe in quantum mechanics. Available: https://physicsworld.com/a/how-the-stern-gerlach-experiment-made-physicists-believe-in-quantum-mechanics/
3. A. Becker, Quantum particles aren't spinning. so where does their spin come from?. Available: https://www.scientificamerican.com/article/quantum-particles-arent-spinning-so-where-does-their-spin-come-from/

8 Photon Polarization

We now repeat the exercise in the previous chapter, applied to photons. By studying a different physical system using the quantum mechanical framework, we aim to strengthen the reader's intuition.

Let's investigate the interaction of light and polarizing filters through a series of experiments (assuming perfect polarizing filters).

Experiment 1

In this experiment, we investigate light from an unpolarized source, e.g., sunlight or an incandescent light bulb. By passing this light through a polarizing light filter aligned horizontally, we observe that the intensity of light has been halved, as illustrated in Fig. 8.1.

Experiment 2

Now we add another polarizing light filter, aligned vertically. Unsurprisingly, we observe experimentally that all the light is blocked as depicted in Fig. 8.2.

Experiment 3

Let's add one more polarizing filter, aligned diagonally at 45 degrees to either the horizontal or vertical polarizer. Furthermore, we can position this filter either first or last. As expected, all of the light remains blocked as shown in Fig. 8.3.

O. M. Raisuddin and S. De, *Quantum Computing for Engineers*,
https://doi.org/10.1007/978-3-032-03325-3_8

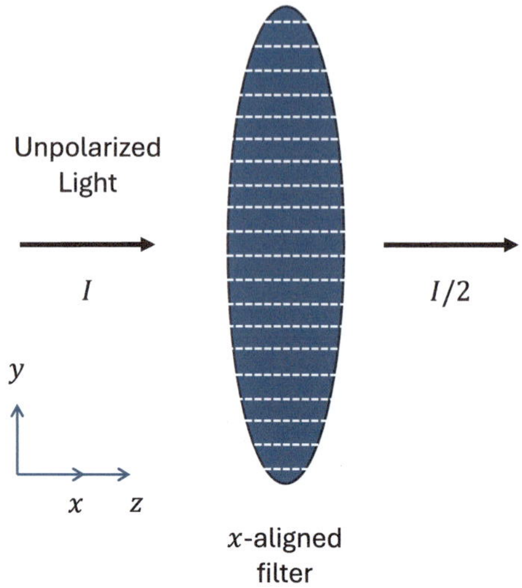

Fig. 8.1 Unpolarized light enters a polarizing filter, and polarized light exits

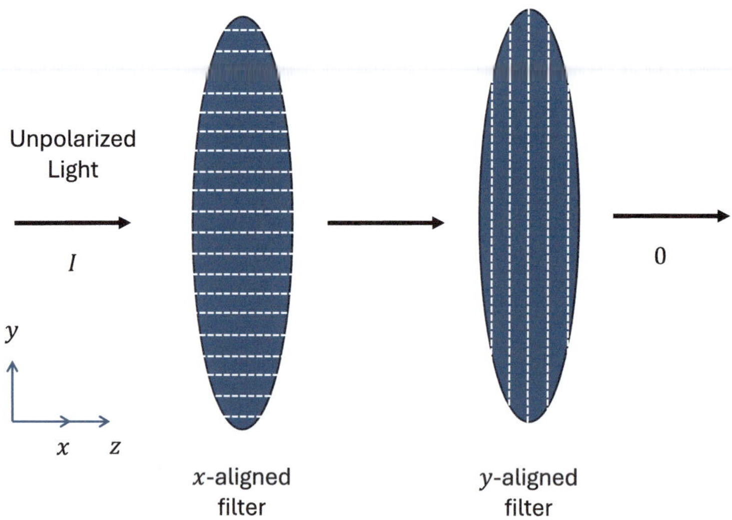

Fig. 8.2 Polarized light is fully filtered by an orthogonal filter

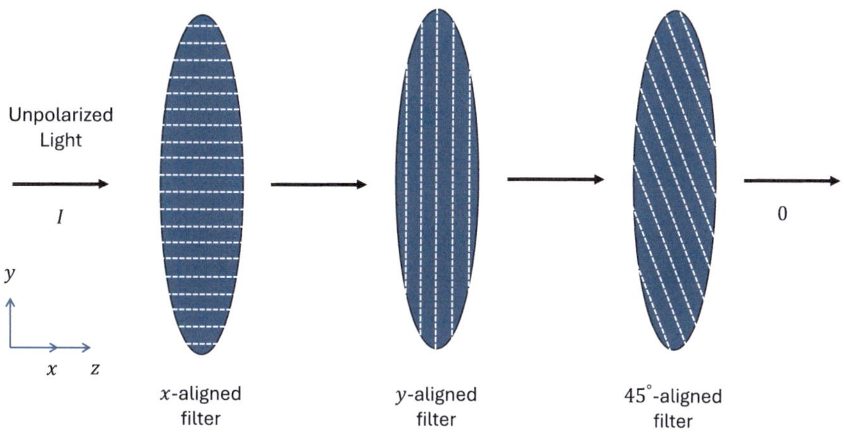

Fig. 8.3 Adding another filter after orthogonally aligned filters yields no change

Experiment 4

Now, let's position the diagonally aligned polarizing filter between the horizontal and vertical filters. Against our classical intuition, we can experimentally observe that some light does pass through the filters. Figure 8.4 shows a polarizing filter rotated 45° placed between the horizontally and vertically aligned filters.

We can now examine these experiments using the quantum mechanical framework. As we did for the Stern–Gerlach experiments, let's assign an orthonormal basis $\{|x\rangle, |y\rangle\}$ to photons polarized horizontally and vertically. Notice that, unlike the basis $\{|+z\rangle, |-z\rangle\}$ used for the Stern–Gerlach experiments, x and y are in fact orthogonal in Euclidean space. We point this out to emphasize that labels used in kets and bras are simply for convenience and may not bear any resemblance to their usual classical interpretation.

Using the $\{|x, |y\rangle\}$ basis, we can model the photons from the light source as

$$|\psi\rangle = \frac{1}{\sqrt{2}}|x\rangle + \frac{1}{\sqrt{2}}e^{i\phi}|y\rangle$$

up to an overall phase where ϕ is an unknown random phase. We can also model horizontally and vertically aligned polarizing filters as projection matrices:

$$F_{hor} = |x\rangle\langle x\rangle = \begin{pmatrix} 1 & 0 \\ 0 & 0 \end{pmatrix}$$

$$F_{ver} = |y\rangle\langle y| = \begin{pmatrix} 0 & 0 \\ 0 & 1 \end{pmatrix}$$

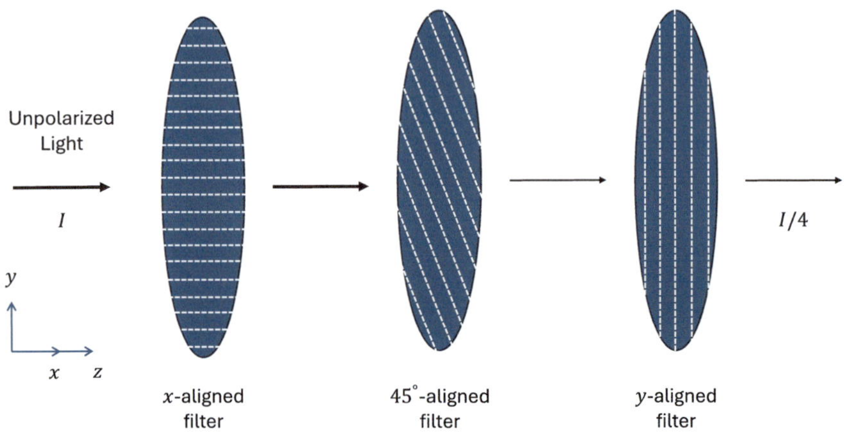

Fig. 8.4 Inserting a filter between orthogonal filters at 45° to both filters allows light to pass through

Note that since photons that are not aligned with the filter are absorbed and/or reflected by the filter, unlike the Stern–Gerlach device, which simply separates the two orthogonal states and allows all particles to pass through.

Furthermore, a diagonally aligned filter can be represented as a rotation of a horizontal filter using a rotation matrix with $\theta = 45°$

$$R_\theta = \begin{pmatrix} \cos(\theta) & -\sin(\theta) \\ \sin(\theta) & \cos(\theta) \end{pmatrix}$$

$$R_{45°} = \frac{1}{\sqrt{2}}\begin{pmatrix} 1 & 1 \\ 1 & -1 \end{pmatrix}$$

$$F_{diag} = R_{45°}F_{hor} = \frac{1}{\sqrt{2}}\begin{pmatrix} 1 & 1 \\ 1 & -1 \end{pmatrix}\begin{pmatrix} 1 & 0 \\ 0 & 0 \end{pmatrix} = \frac{1}{\sqrt{2}}\begin{pmatrix} 1 & 0 \\ 1 & 0 \end{pmatrix}$$

By passing unpolarized photons through a horizontally aligned filter, according to Experiment 1, we get

$$\langle\psi|F_{hor}|\psi\rangle = \frac{1}{2}|x\rangle$$

which matches the 50% light intensity observed experimentally. Introducing a vertically aligned filter as laid out in Experiment 2, we get

$$\langle\psi|F_{ver}F_{hor}|\psi\rangle = \langle\psi|\begin{pmatrix} 0 & 0 \\ 0 & 0 \end{pmatrix}|\psi\rangle = 0$$

as expected.

For Experiment 3, we can clearly see that

$$\langle\psi|F_{diag}F_{ver}F_{hor}|\psi\rangle = \langle\psi|F_{diag}\begin{pmatrix}0 & 0\\ 0 & 0\end{pmatrix}|\psi\rangle = \langle\psi|\begin{pmatrix}0 & 0\\ 0 & 0\end{pmatrix}|\psi\rangle = 0$$

Finally, for Experiment 4, we can compute

$$\langle\psi|F_{ver}F_{diag}F_{hor}|\psi\rangle = \frac{1}{4}\begin{pmatrix}1 & 1\end{pmatrix}\begin{pmatrix}0 & 0\\ 0 & 1\end{pmatrix}\begin{pmatrix}1 & 0\\ 1 & 0\end{pmatrix}\begin{pmatrix}1 & 0\\ 0 & 0\end{pmatrix}\begin{pmatrix}1\\ 1\end{pmatrix} = \frac{1}{4}$$

matching the 25% light intensity observed experimentally.

In the discussion above, we have barely scratched the surface of the rich physics modeled by a quantum mechanical treatment of photons. As an example, one may also create basis vectors

$$|R\rangle = \frac{1}{\sqrt{2}}|x\rangle + \frac{1}{\sqrt{2}}i|y\rangle, \, |L\rangle = \frac{1}{\sqrt{2}}|x\rangle - \frac{1}{\sqrt{2}}i|y\rangle$$

to represent right- and left-circularly polarized light. However, these discussions are beyond the scope of this text.

Part III
The Quantum Computing Model

Armed with the mathematical preliminaries and having seen a preview of the non-intuitive nature of quantum mechanics, we now begin to formally introduce the quantum computing model. We emphasize clarity and functional understanding over exhaustive technical detail. Readers are guided through qubits, quantum gates, measurements, and circuit-based models—concepts that will recur throughout the remainder of the book. Familiarity with these elements is essential for understanding how quantum algorithms are constructed and analyzed.

Our discussion starts directly with the quantum circuit model. Historically, quantum Turing machines were the first quantum computational model. Since quantum Turing machines are unwieldy and non-intuitive, the ideas behind universal circuit families have been applied to formulate the quantum circuit model, which is polynomially equivalent to quantum Turing machines.

To motivate the practical realization of quantum computers, we begin by briefly reviewing the DiVincenzo criteria, which need to be fulfilled to build a quantum computer [1]:

1. A physical machine consisting of a scalable number of qubits.

2. The ability to initialize the qubits in a known quantum state.

3. The qubits must have coherence times longer than the time required to execute a quantum gate.

4. The ability to perform a universal set of gates.

5. The ability to measure qubits.

The chapters in this part provide foundational knowledge and build progressively toward more complex quantum computing concepts:

Chapter 9, "Qubits, Quantum Registers, and Quantum Gates", introduces qubits—the fundamental units of quantum information—and describes how multiple qubits combine into quantum registers. Essential quantum gates and their algebraic properties are presented.

Chapter 10, "Quantum Measurements and Circuits", covers quantum measurement theory, introducing measurement operators and the quantum circuit model. It also addresses practical aspects like bitstring sampling and the principle of deferred measurement.

Chapter 11, "Superposition and Entanglement", explores the uniquely quantum concepts of superposition and entanglement using quantum circuits, emphasizing intuition and illustrative examples.

Chapter 12, "Classical and Reversible Computation", bridges classical computation and quantum logic, explaining classical logic embedding into quantum circuits and discussing the concept of reversible computation and quantum oracles.

Chapter 13, "Access Models and Data Representation", introduces quantum access models, including sparse access and block-encoding models. It also discusses Hermitian dilation and Pauli-basis decomposition, crucial for understanding advanced quantum algorithms.

Chapter 14, "Limitations of Quantum Computers", discusses fundamental theoretical limits on quantum computing, including key no-go theorems such as the no-cloning and no-deletion theorems, and limitations on quantum speedups.

Chapter 15, "Simon's, Deutsch–Jozsa, and Bernstein–Vazirani Algorithms", provides concrete examples of quantum algorithms demonstrating exponential speedups compared to classical counterparts, emphasizing complexity class separations and the potential of quantum computing. The Abelian hidden subgroup problem is introduced as a unifying framework for exponential speedups.

Reference

1. D.P. DiVincenzo, D. Loss, Quantum information is physical. Superlattices and Microstructures. **23**(3–4), 419–432 (1998). https://doi.org/10.1006/spmi.1997.0520

9 Qubits, Quantum Registers, and Quantum Gates

Qubits

Classical digital computers use binary "bits," 0 and 1, as fundamental units of information. Quantum computers use an analogous quantum bit, or a "qubit," as their elementary unit. While a classical bit can exist in only one of two definite states, a qubit can exist in a superposition of both $|0\rangle$ and $|1\rangle$, enabling fundamentally new modes of computation.

Definition:
A qubit (quantum bit) is a quantum system whose state is represented by a unit vector $|\psi\rangle$ in a 2D complex Hilbert space $\mathcal{H} \cong \mathbb{C}^2$. The state can be expressed as a linear combination (superposition) of two orthogonal basis states, typically denoted as $|0\rangle$ and $|1\rangle$, such that

$$|\psi\rangle = \alpha|0\rangle + \beta|1\rangle$$

where $\alpha, \beta \in \mathbb{C}$ and $|\alpha|^2 + |\beta|^2 = 1$. $|\alpha|^2, |\beta|^2$ are the probabilities of measuring the qubit in the states $|0\rangle, |1\rangle$, respectively.

Recall that $|0\rangle$ and $|1\rangle$ are orthonormal basis states in the $0 - 1$ basis, also referred to as the Z basis, which can be represented as the basis vectors $|0\rangle = \begin{pmatrix} 1 \\ 0 \end{pmatrix}$ and $|1\rangle = \begin{pmatrix} 0 \\ 1 \end{pmatrix}$.

Alternative orthonormal bases are often used to highlight different measurement contexts:

O. M. Raisuddin and S. De, *Quantum Computing for Engineers*,
https://doi.org/10.1007/978-3-032-03325-3_9

- The X basis consists of the states $|+\rangle$ and $|-\rangle$, defined as

$$|+\rangle = \frac{1}{\sqrt{2}}(|0\rangle + |1\rangle), |-\rangle = \frac{1}{\sqrt{2}}(|0\rangle - |1\rangle)$$

- The Y basis consists of the states $|+i\rangle$ and $|-i\rangle$, defined as

$$|+i\rangle = \frac{1}{\sqrt{2}}(|0\rangle + i|1\rangle), |-i\rangle = \frac{1}{\sqrt{2}}(|0\rangle - i|1\rangle)$$

The state of a qubit may be visualized as a point on the surface of a "Bloch sphere," as shown in Fig. 9.1. However, this visualization omits the global phase of a qubit (e.g., a quantum state $|\psi\rangle = -|0\rangle$ or $|\psi\rangle = i|0\rangle$ appears identical on the Bloch sphere) and does not readily generalize to multi-qubit systems. However, it is a useful visual tool to understand the relation between various qubit bases and the action of single-qubit gates.

A qubit may be measured in an arbitrary (orthogonal) basis $|\psi\rangle = \alpha|I\rangle + \beta|II\rangle$. According to the measurement postulate (Part II: A Brief Introduction to Quantum Mechanics) the basis states $|I\rangle$ and $|II\rangle$ can be observed with probabilities $|\alpha|^2$ and $|\beta|^2$, respectively. If no operations are applied after measurement, further measurements in the same basis ($|I\rangle$ and $|II\rangle$) will repeatedly yield the same qubit state that was originally measured.

Physically, a qubit is typically realized using a two-state quantum system (e.g., the electronic states of ions) or by utilizing two energy levels of a quantum system (e.g., the ground state and first excited state of a quantum harmonic oscillator).

The keen reader would automatically inquire about the implications of other energy levels. Indeed, they can be utilized, and this conceptual extension is known

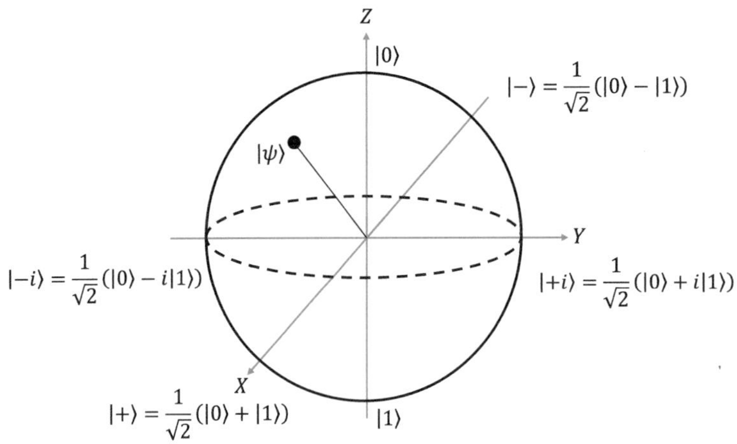

Fig. 9.1 A Bloch sphere with the X, Y, and Z bases labeled, and an arbitrary quantum state $|\psi\rangle$ on the surface of the sphere

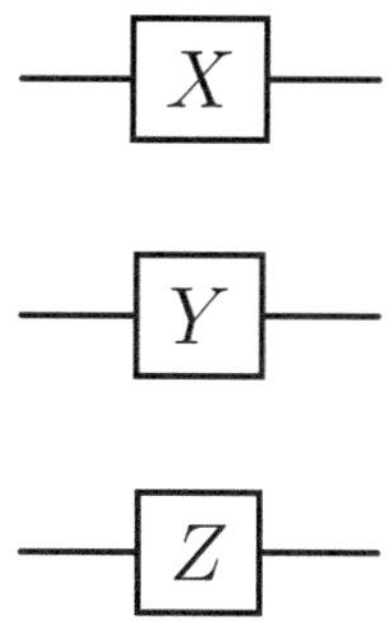

Fig. 9.2 Single-qubit X, Y, and Z gates operating on qubits

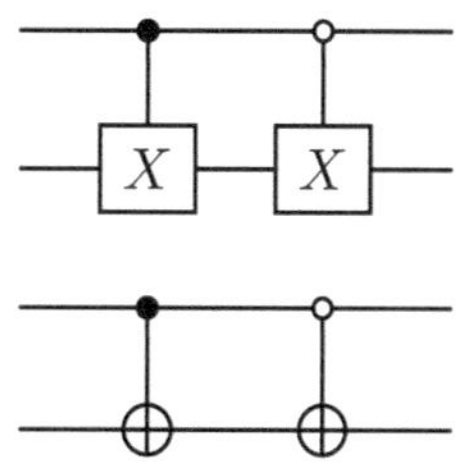

Fig. 9.3 Controlled single-qubit cX (or CNOT) gates. The gates on the left are conditioned on the control qubit being in the $|1\rangle$ state, indicated by a filled circle. The representations at the top and bottom are equivalent

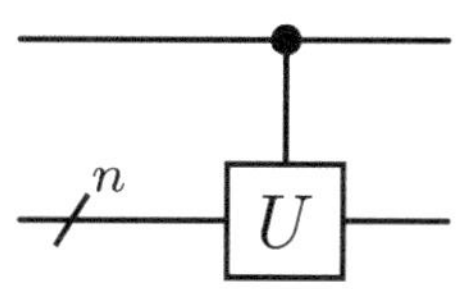

Fig. 9.4 A multi-qubit controlled gate

as “qudits.” However, since qudits are not used in this book (and can be mapped onto qubits), we burden the interested reader with seeking out this information.

Registers of Qubits

Similar to a classical register of bits, a set of qubits may be combined into a register of qubits. However, the state of a register of qubits differs significantly from a classical register of bits.

Definition: A register of n qubits is represented as a state vector $|\psi\rangle \in \mathcal{H}$ where $\mathcal{H} \cong \mathbb{C}^{2^n}$ and is described as a superposition of 2^n orthogonal states $|0\rangle, |1\rangle, \ldots, |2^n - 1\rangle$ as $|\psi\rangle = \sum_{i=0}^{2^n-1} \alpha_i |i\rangle$ where the probability amplitudes α_j satisfy the probability completeness relation $\sum_{i=0}^{2^n-1} |\alpha_i|^2 = 1$ according to the Born rule.

A quantum register is a tuple of n qubits, whose combined state is represented as the quantum system $|\psi\rangle \in \mathbb{C}^{2^n}$, corresponding to a tensor product of the individual Hilbert spaces of the qubits.

The combined state $|\psi\rangle$ of the two qubits $|\psi_1\rangle = \alpha_1|0\rangle + \beta_1|1\rangle$ and $|\psi_2\rangle = \alpha_2|0\rangle+\beta_2|1\rangle$ may be represented in Dirac notation as a Kronecker product, denoted by $\otimes$, of the individual qubits with various equivalent notations:

$$\begin{aligned}
|\psi\rangle &= |\psi_1\rangle \otimes |\psi_2\rangle = |\psi_1\rangle|\psi_2\rangle = |\psi_1\psi_2\rangle \\
&= (\alpha_1|0\rangle + \beta_1|1\rangle) \otimes (\alpha_2|0\rangle + \beta_2|1\rangle) \\
&= \alpha_1\alpha_2|0\rangle|0\rangle + \alpha_1\beta_2|0\rangle|1\rangle + \alpha_2\beta_1|1\rangle|0\rangle + \beta_1\beta_2|1\rangle|1\rangle \\
&= \alpha_1\alpha_2|00\rangle + \alpha_1\beta_2|01\rangle + \alpha_2\beta_1|10\rangle + \beta_1\beta_2|11\rangle \\
&= \alpha_1\alpha_2\begin{pmatrix}1\\0\end{pmatrix} \otimes \begin{pmatrix}1\\0\end{pmatrix} \\
&+ \alpha_2\beta_1\begin{pmatrix}1\\0\end{pmatrix} \otimes \begin{pmatrix}0\\1\end{pmatrix} + \alpha_1\beta_2\begin{pmatrix}0\\1\end{pmatrix} \otimes \begin{pmatrix}1\\0\end{pmatrix} + \beta_1\beta_2\begin{pmatrix}0\\1\end{pmatrix} \otimes \begin{pmatrix}0\\1\end{pmatrix} \\
&= \alpha_1\alpha_2\begin{pmatrix}1\\0\\0\\0\end{pmatrix} + \alpha_2\beta_1\begin{pmatrix}0\\1\\0\\0\end{pmatrix} + \alpha_1\beta_2\begin{pmatrix}0\\0\\1\\0\end{pmatrix} + \beta_1\beta_2\begin{pmatrix}0\\0\\0\\1\end{pmatrix}
\end{aligned}$$

and in vector form, using the computational basis:

$$|\psi\rangle = \begin{pmatrix}\alpha_1\alpha_2\\ \alpha_2\beta_1\\ \alpha_1\beta_2\\ \beta_1\beta_2\end{pmatrix}$$

Additional qubits will follow the same pattern, resulting in an exponentially large state space for the qubit register. A change in the order of the qubits in the quantum register simply shuffles the representation of the state corresponding to the definition of the Kronecker product. Registers of qubits can be used to represent data in various formats, which is discussed in Chapter 13: Access Models and Data Representation.

Analogous to logic gates, which operate on classical bits and registers of bits, quantum gates operate on qubits and registers of qubits. While classical logic gates take one or more classical bits as input and produce classical output states deterministically or probabilistically based on the input, a quantum gate acts on one or more qubits and outputs the same qubits in a new quantum state, generally via a unitary transformation.

Unlike classical logical operations, where the input bits remain available after the application of a logic gate, the input state of a quantum gate is not readily available after application of a gate. By the principles of quantum mechanics, quantum gates are unitary operators. Thus, it is always possible to reverse a quantum gate to recover the input state, though this comes at the expense of losing access to the output state. In contrast, most classical logic gates (e.g., AND, OR) are not reversible; their inputs cannot, in general, be reconstructed from the outputs.

Quantum Gates

In the gate-based quantum computing model, operations on qubits are represented as quantum gates, or simply gates. Some quantum gates are analogous to classical gate operations, e.g., the X gate is analogous to a classical *NOT* operation. However, quantum gates may not have a corresponding classical counterpart, e.g., a Hadamard H gate.

Definition: A quantum gate operation on n qubits is represented as a unitary matrix $U \in \mathbb{C}^{2^n \times 2^n}$.

Quantum gates can be conveniently represented as complex unitary matrices, in line with the column vector representation of qubits. The simplest gates are single-qubit gates represented as $SU(2) \in \mathbb{C}^{2\times 2}$ matrices. Some important single-qubit gates are the identity gate I, the Pauli gate set $\{X, Y, Z\}$, and the Hadamard gate H. Their matrix representations and descriptions are provided in Table. 9.1.

As an example, the X gate is represented as the unitary matrix

$$X = \begin{pmatrix} 0 & 1 \\ 1 & 0 \end{pmatrix}$$

The X gate is the quantum analog of the classical NOT gate. As an example, applying the X gate to the basis state $|0\rangle$ yields $|1\rangle$ and vice versa:

$$X|0\rangle = \begin{pmatrix} 0 & 1 \\ 1 & 0 \end{pmatrix}\begin{pmatrix} 1 \\ 0 \end{pmatrix} = \begin{pmatrix} 0 \\ 1 \end{pmatrix} = |1\rangle$$

$$X|1\rangle = \begin{pmatrix} 0 & 1 \\ 1 & 0 \end{pmatrix}\begin{pmatrix} 0 \\ 1 \end{pmatrix} = \begin{pmatrix} 1 \\ 0 \end{pmatrix} = |0\rangle$$

and in general

$$X|\psi\rangle = X(\alpha|0\rangle + \beta|1\rangle) = \beta|0\rangle + \alpha|1\rangle$$

Table. 9.1 lists various commonly used gates with their matrix representations and descriptions.

Quantum gates can also act on n qubits, in which case they can be represented in $SU(2^n) \in \mathbb{C}^{2^n \times 2^n}$. Multiple qubit gates typically arise as controlled versions of gates, of which *cX* (or *cNOT*) is a commonly used one. A controlled gate manipulates the state of a "target qubit," conditioned on the state of a "control qubit." The *SWAP* gate is another common gate that swaps states between qubits. Controlled gates can be represented in block form. As an example, the matrix

Table 9.1 Common gates and their descriptions

Gate	Matrix representation	Eigendecomposition	Description
I	$\begin{pmatrix} 1 & 0 \\ 0 & 1 \end{pmatrix}$	–	No operation on qubit. Qubit state is unchanged
X	$\begin{pmatrix} 0 & 1 \\ 1 & 0 \end{pmatrix}$	$\frac{1}{\sqrt{2}}\begin{pmatrix} 1 & 1 \\ 1 & -1 \end{pmatrix}\begin{pmatrix} 1 & 0 \\ 0 & -1 \end{pmatrix}\frac{1}{\sqrt{2}}\begin{pmatrix} 1 & 1 \\ 1 & -1 \end{pmatrix}$ HZH	Pauli X gate and matrix. 180° rotation around X. Transform basis states $\|0$ to $\|1$ and $\|1$ to $\|0$
Y	$\begin{pmatrix} 0 & -i \\ i & 0 \end{pmatrix}$	$\left[\begin{pmatrix} 1 & 0 \\ 0 & i \end{pmatrix}\frac{1}{\sqrt{2}}\begin{pmatrix} 1 & 1 \\ 1 & -1 \end{pmatrix}\right]\begin{pmatrix} 1 & 0 \\ 0 & -1 \end{pmatrix}\left[\frac{1}{\sqrt{2}}\begin{pmatrix} 1 & 1 \\ 1 & -1 \end{pmatrix}\begin{pmatrix} 1 & 0 \\ 0 & -i \end{pmatrix}\right]$ $SHZHS^{\dagger}$	Pauli Y gate and matrix. 180° rotation around Y
Z	$\begin{pmatrix} 1 & 0 \\ 0 & -1 \end{pmatrix}$	$\begin{pmatrix} 1 & 0 \\ 0 & -1 \end{pmatrix}$ Z	Pauli Z gate and matrix.180° rotation around Z. Apply phase of -1 to $\|1$
H	$\frac{1}{\sqrt{2}}\begin{pmatrix} 1 & 1 \\ 1 & -1 \end{pmatrix}$	–	90° CW rotation around X. Used to create superposition from basis state $\|0$
S	$\begin{pmatrix} 1 & 0 \\ 0 & i \end{pmatrix}$	–	Phase gate. Equivalent to $\sqrt{Z}$ gate. $90°$ rotation around Z. Apply phase of i to $\|1$

Note that the eigenstates of the X, Y, and Z gates coincide with the X, Y, and Z bases shown on the Bloch sphere in Fig. 9.1. This is where the terminology "measuring in the X, Y, or Z" basis comes from. This is discussed in more detail in Chapter 10: Quantum Measurements and Circuits

representation of the cX gate, where the first qubit from the left is the control qubit and the second qubit is the target qubit, is

$$cX = |0\rangle\langle 0| \otimes I + |1\rangle\langle 1| \otimes X = \begin{pmatrix} I & \\ & X \end{pmatrix} = \begin{pmatrix} 1 & 0 & & \\ 0 & 1 & & \\ & & 0 & 1 \\ & & 1 & 0 \end{pmatrix}$$

where $|\cdot\rangle\langle\cdot|$ is an outer product.

Controlled gates may be conditioned on the $|1\rangle$ state as well. A cX gate conditioned on $|1\rangle$ operates as

$$|0\rangle\langle 0| \otimes X + |1\rangle\langle 1| \otimes I = \begin{pmatrix} X & \\ & I \end{pmatrix} = \begin{pmatrix} 0 & 1 & & \\ 1 & 0 & & \\ & & 1 & 0 \\ & & 0 & 1 \end{pmatrix}$$

In general, controlled gates may be conditioned on multiple qubits and may apply any general unitary operation. A general unitary gate U conditioned on the logical state of k qubits represented as a bitstring bin(i), where the bitstring is formed from a binary representation of $\{i \in \mathbb{Z} |\ i \geq 0\}$, operates as

$$c_k U = \sum_{j \neq i} |j\rangle\langle j| \otimes I + |i\rangle\langle i| \otimes U$$

and can be represented as a block-diagonal matrix with the matrix representation of U on the ith block and I on the remaining blocks:

$$c_k U = \begin{pmatrix} I & & & & & & \\ & \ddots & & & & & \\ & & I & & & & \\ & & & U & & & \\ & & & & I & & \\ & & & & & \ddots & \\ & & & & & & I \end{pmatrix}$$

All quantum gates are unitary operations, which ensures the normalization of quantum states according to the Born rule. Consequently, all quantum gates are also reversible operations, with the reverse operation simply being the Hermitian transpose or conjugate transpose of the quantum gate.

Gates acting on qubits correspond to left multiplication, with the ket representing the quantum state of the qubits. As an example, consider a register of three qubits in the state $|000\rangle$. Applying a Hadamard gate to the first qubit (from the left) and a Pauli X gate to the last qubit is represented as

$$(H \otimes I \otimes X)|000\rangle = (H \otimes I \otimes I)|001\rangle = \frac{1}{\sqrt{2}}(|001\rangle + |101\rangle)$$

As another example, consider a register of two qubits in the state $|00\rangle$ with a Hadamard gate applied to the first qubit followed by a *CNOT* gate controlled by the first qubit applied to the second qubit:

$$(|00\rangle\langle 00| + |01\rangle\langle 01| + |10\rangle\langle 11| + |11\rangle\langle 10|)|00\rangle = |00\rangle + |11\rangle$$

Note that the order of operations progresses from right to left as usual for matrix multiplication, which is the opposite of the order used for quantum circuits introduced in Chap. 10: Quantum Measurements and Circuits. It is customary not to include the I gate when the qubits on which the operation is applied are implied.

Quantum gates are one of the building blocks of quantum circuits (discussed in Chap. 10: Quantum Measurements and Circuits). They are represented in quantum circuits as boxes placed on lines representing the qubits they act on. The label inside the box indicates the type of gate. Multiple quantum gates may also be combined in this representation with a description for brevity when describing algorithms. The control bits in controlled gates are represented as filled or empty circles, depending on whether the gate is conditioned on that qubit being in the state $|0\rangle$ or $|1\rangle$ respectively.

Like classical Boolean logic gates, quantum gates can also form a universal gate set. The fundamental difference is that while classical universal gate sets can implement arbitrary logical operations exactly, quantum universal gate sets can approximate arbitrary unitary operations to arbitrary precision.

This implies that any arbitrary quantum gate can be approximated using a universal gate set. The Solovay–Kitaev theorem [1] is a central theorem in quantum computing which shows that the approximation error using a universal gate set scales as $O\left(\log^c\left(\frac{1}{\epsilon}\right)\right)$ for a single-qubit gate where $c \approx 2$ and $O\left(m \log^c\left(\frac{m}{\epsilon}\right)\right)$ for a set of *mCNOT* s and single-qubit unitaries. This corresponds to a polylogarithmic increase in the approximation using a universal gate set over the original number of arbitrary gates, which is efficient. The statement of the theorem for qubits is as follows.

Theorem [2]: Let $\mathcal{G}$ be an instruction set for $SU(d)$, and let a desired accuracy $\epsilon > 0$ be given. There is a constant c such that for any $U \in SU(d)$ there exists a finite sequence S of gates from $\mathcal{G}$ of length $O(\log^c(1/\epsilon))$ and such that $U - S \leq \epsilon$.

The physical interpretation of a quantum gate depends on the underlying architecture. For a superconducting transmon qubit architecture, quantum gates are typically implemented as microwave pulses corresponding to the resonant frequency of the transmon qubits. Underlying hardware implementations can have a variety of gate sets. Algorithms are typically agnostic to the hardware implementation since all quantum gates are converted to the target hardware's gate set (implemented as physical processes) in a process called transpilation. We discuss the quantum computer programming stack in more detail in Part IV: Programming Quantum Computers. The underlying hardware may also have a specific connectivity graph, which dictates the possibility of controlled gate operations between qubits.

The *SWAP* gate is another common gate that swaps the states between qubits, i.e., performs the map $|\psi\rangle|\phi\rangle \rightarrow |\phi\rangle|\psi\rangle$. While this may seem trivial as interchanging the order of qubits, it is typically necessary for applied quantum computing on realistic hardware which does not have all-to-all connectivity for two-qubit interactions between qubits.

The transpilation process introduces *SWAP* gates to "move" qubit states around to allow controlled operations between any set of qubits that might not be connected by the topology of the target hardware. Although this is not an issue for "fault-tolerant" or "error-corrected" devices, a multitude of *SWAP* gates can introduce unwanted noise into the system in current realistic devices, making them a point of concern for NISQ hardware. Similarly, controlled gates (or multi-controlled) gates typically introduce errors larger than those of single-qubit gates.

References

1. A. Y. Kitaev, Quantum computations: algorithms and error correction, Russ. Math. Surv., **52**(16), pp. 1191–1249, (1997). https://doi.org/10.1070/RM1997v052n06ABEH002155
2. C. M. Dawson, M. A. Nielsen, The Solovay-Kitaev algorithm. Quantum Info. Comput., **6**(1), pp. 81–95, (2006)

Quantum Measurements and Circuits 10

In this chapter, we introduce the measurement operation to read out the states of qubits. Later in the chapter, we combine registers of qubits, gates, and measurements to form quantum circuits.

A quantum state is defined using probability amplitudes. To read a state, a series of measurements of the state needs to be performed, and the measurement statistics will correspond to the probability amplitudes of the quantum state and the basis used for the measurement [1]. Note that, in general, to exactly infer the probability amplitudes of a quantum state through measurements, an infinite number of measurements is needed in accordance with the Law of Large Numbers, regardless of the number of qubits.

Measurement Operators

We first define a measurement operator M_m for measuring a quantum state $|\psi\rangle$.

Definition: Given a quantum state $|\psi\rangle$ and a measurement operator M_m where m is a measurement outcome, the probability of measuring an outcome m is.

$$p(m) = \langle\psi|M_m^\dagger M_m|\psi\rangle$$

and the state of the quantum system after measuring an outcome m is

$$\frac{M_m|\psi\rangle}{\sqrt{p(m)}}$$

such that the completeness relation

$$\sum_m p(m) = 1$$

O. M. Raisuddin and S. De, *Quantum Computing for Engineers*,
https://doi.org/10.1007/978-3-032-03325-3_10

Fig. 10.1 Symbol for a measurement operation

is satisfied.

Measurement operators are projection operations. In gate-based quantum computing, projective measurements in the computational basis (the Z or $|0\rangle$-$|1\rangle$ basis) are typically used corresponding to

$$M_0 = |0\rangle\langle 0| = \begin{pmatrix} 1 & 0 \\ 0 & 0 \end{pmatrix}, M_1 = |1\rangle\langle 1| = \begin{pmatrix} 0 & 0 \\ 0 & 1 \end{pmatrix}$$

and are represented by a "meter" symbol in quantum circuits (Fig. 10.1).

For measurements over multiple qubits, we use the notation M_j where j is an integer or bitstring indicating the measurement outcome. In quantum algorithms, measurements may be performed over a few qubits in registers or all qubits.

As an example, consider a two-qubit system in the state

$$|\psi\rangle = \frac{1}{2}(|00\rangle + |01\rangle + |10\rangle + |11\rangle)$$

with a measurement being performed on only the first qubit. The first qubit is measured in the state $|0\rangle$ with

$$\begin{aligned} p(0) &= \langle\psi|M_0^\dagger M_0|\psi\rangle = \langle\psi|M_0|\psi\rangle \\ &= \left(\frac{1}{2}(|00\rangle + |01\rangle + |10\rangle + |11\rangle)\right)^\dagger (|0\rangle\langle 0| \otimes I)\left(\frac{1}{2}(|00\rangle + |01\rangle + |10\rangle + |11\rangle)\right) = \frac{1}{2} \end{aligned}$$

and the final state after measuring the first qubit in the state $|0\rangle$ is

$$\begin{aligned} \frac{M_0|\psi\rangle}{\sqrt{p(0)}} &= \frac{(|0\rangle\langle 0| \otimes I)\left(\frac{1}{2}(|00\rangle + |01\rangle + |10\rangle + |11\rangle)\right)}{\frac{1}{\sqrt{2}}} \\ &= \frac{1}{\sqrt{2}}(|00\rangle + |01\rangle) \end{aligned}$$

Note the renormalization of the system according to the Born rule [2]. Similarly, the first qubit is measured in the state $|1\rangle$ with $p(1) = \frac{1}{2}$ and the final state is

$$\begin{aligned} \frac{M_1|\psi\rangle}{\sqrt{p(1)}} &= \frac{(|1\rangle\langle 1| \otimes I)\left(\frac{1}{2}(|00\rangle + |01\rangle + |10\rangle + |11\rangle)\right)}{\frac{1}{\sqrt{2}}} \\ &= \frac{1}{\sqrt{2}}(|10\rangle + |11\rangle) \end{aligned}$$

Measurements may be performed to either read out an entire quantum register or as a flag to indicate the successful completion of an operation. Qubits used as a flag are typically called ancilla qubits, and their measurement can signal whether a desired computational step has succeeded or requires repetition.

Note that measurements will uncover the squared moduli (probability) of the complex numbers (probability amplitudes) defining the probability amplitude of a quantum state. To recover additional information, e.g., the phases of the probability amplitudes, a process known as quantum state tomography [3] is performed.

Quantum states also have an overall phase, which is not measurable; only a relative overall phase between two states is measurable. As an example, consider the two quantum states

$$|\psi\rangle = \alpha|0\rangle + \beta|1\rangle, |\phi\rangle = i(\alpha|0\rangle + \beta|1\rangle)$$

The phase i distinguishes $|\psi\rangle$ from $|\phi\rangle$ and cannot be detected from projective measurements alone, since

$$\langle\psi\left|M_0^\dagger M_0\right|\psi\rangle = \langle\phi\left|M_0^\dagger M_0\right|\phi\rangle = |\alpha|^2$$
$$\langle\psi\left|M_1^\dagger M_1\right|\psi\rangle = \langle\phi\left|M_1^\dagger M_1\right|\phi\rangle = |\beta|^2$$

A hardware implementation will typically only allow a Z basis ($|0\rangle$-$|1\rangle$) measurement. To perform measurements in another basis (e.g., X or Y bases), the qubits can be "rotated" to the desired basis by applying gate operations and then measured. This is discussed later in this chapter.

Bitstring Sampling

Bitstring sampling is straightforward: circuits are executed repeatedly, and qubit registers are measured at the end of each execution. The sampled bitstrings are then used for further computation. Sampling a quantum state $|\psi\rangle \in \mathbb{C}^{2^n}$ can be described mathematically as sampling from the distribution

$$p(j) = \langle\psi|M_j^\dagger M_j|\psi\rangle \ \forall\, j \in \left[0, 2^n - 1\right]$$

which will yield an outcome j with probability $p(j)$. In its raw form, data output from a quantum computer is a distribution of sampled bitstrings.

Quantum Circuits

Working with algebraic forms of algorithms can be unwieldy and difficult to visualize. Quantum circuits are a convenient visual representation of quantum algorithms, clearly showing sequences of gates and measurements acting on qubits. Each horizontal line in a quantum circuit diagram represents a qubit, and each symbol or block placed on these lines represents quantum gates or measurements.

For example, consider a three-qubit register $|ABC\rangle$. Suppose an H gate is applied to qubit A, an X gate is applied to qubit C, followed by a $CNOT$ gate controlled by qubit A acting on qubit B, and, finally, all qubits are measured. These operations can conveniently be represented in a quantum circuit, as shown in Fig. 10.2.

Note that the operations in a quantum circuit are ordered from left to right, unlike algebraic notation, which typically lists operations from right to left. For instance, the sequence of unitary operations $U_A U_B U_C|\psi\rangle$ is represented in a quantum circuit as shown in Fig. 10.3.

Additionally, controlled operations conditioned on the control qubit being either in the state $|1\rangle$ or $|0\rangle$ are represented using filled or empty circles, respectively, as shown in Fig. 10.4.

Quantum registers can be ordered using either little-endian or big-endian notation. In little-endian ordering, the rightmost qubit in algebraic notation is indexed as 0, while in big-endian ordering, the leftmost qubit indexed as 0. In a quantum circuit, the first qubit from the top of a register is indexed as 0. When implementing quantum algorithms, endian consistency is critical; mixing the ordering results in

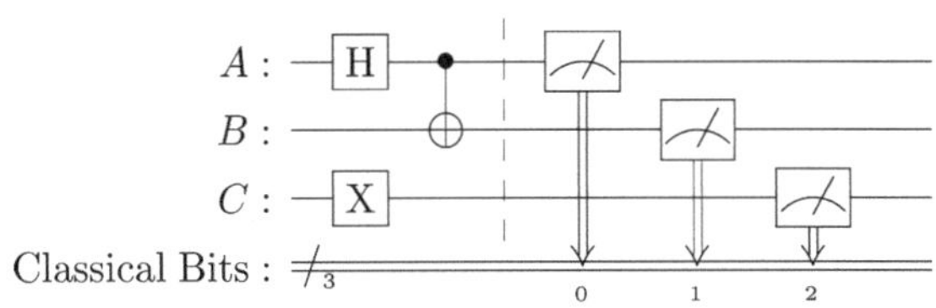

Fig. 10.2 Quantum circuit representation of gate and measurement operations

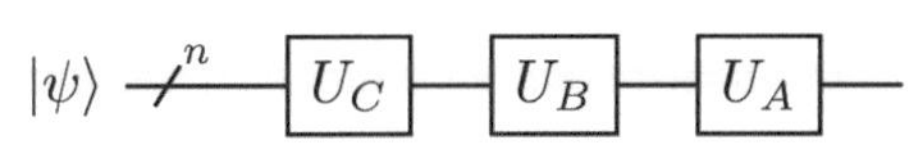

Fig. 10.3 Order of operations in circuit representation for $U_A U_B U_C|\psi\rangle$

Fig. 10.4 Controlled X gates conditioned on Left: $|1\rangle$; Right: $|0\rangle$

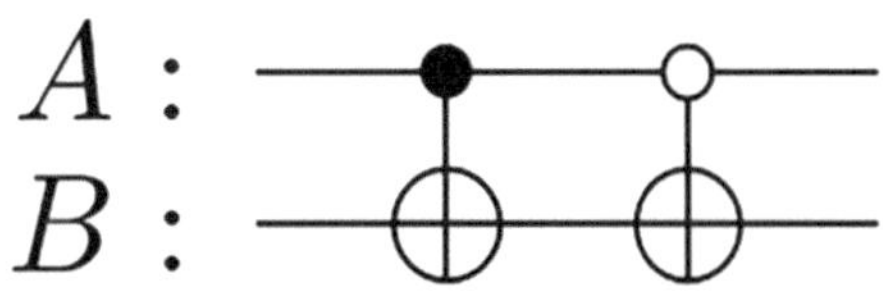

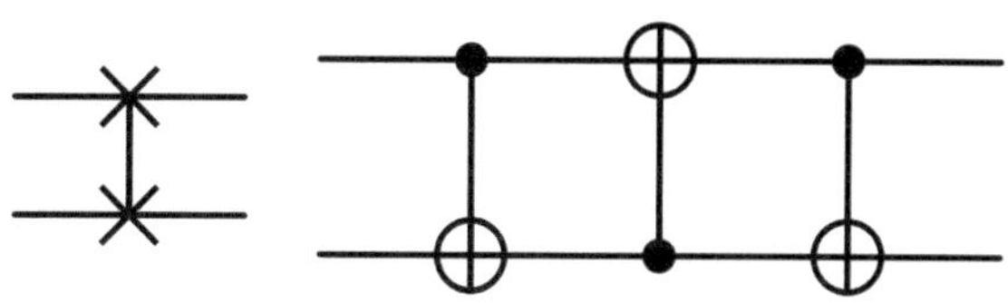

Fig. 10.5 Left: A SWAP gate; Right: Decomposition of a SWAP gate

a shuffling of the basis states. The shuffling leads to the measurement distribution being distorted, leading to an incorrect interpretation of the measurement distribution. Both big- and little-endian conventions are used in libraries for quantum computing. As examples, Qiskit [4] uses little-endian notation, while Pennylane [5] uses big-endian notation. We illustrate this using a register of three qubits $qreg = |q_1q_2q_3\rangle$. In Qiskit (Little endian), qubit q_3 will be indexed classically as $qreg[0]$, and in Pennylane (Big endian) it will be indexed as $qreg[2]$. Additionally, Pauli strings (introduced in Chap. 13: Access Models and Data Representation) must also be indexed consistently. In this book all figures and equations are in big-endian notation, and all Qiskit code is in little-endian notation.

Another important gate we revisit here is the SWAP gate. In quantum computers, operations between arbitrary qubits are not always directly possible due to the physical connectivity of the qubits. However, it is always possible to shuttle around the quantum state of a qubit using SWAP gates. The symbol for a SWAP gate and its decomposition are shown in Fig. 10.5.

Quantum circuits can be viewed as directed acyclic graphs, where the nodes represent gates and the edges represent the qubits on which the gates operate. Quantum circuits may also represent hybrid quantum–classical operations, such as mid-circuit measurements, classical conditional logic, and dynamic circuit structures. However, these advanced circuit concepts are outside the scope of this book and are not used in the remainder of this book.

In practice, both the *circuit depth* (the number of sequential gate layers) and *circuit width* (the number of qubits used) are important resource metrics that affect whether a given circuit can be realistically implemented on current hardware. Specialized algorithms and quantum compilers exist that optimize circuit depth and width, often by reordering gates, identifying redundant operations, or mapping the circuit to the native gate set of the target hardware. Furthermore, since real quantum hardware supports only a limited set of *native gate sets*, quantum circuits must often be *compiled* from high-level descriptions into sequences of hardware-supported gates, which can introduce additional overhead. The effects of *noise* and *decoherence* in current devices make minimizing circuit depth critical for obtaining reliable results. Some platforms now also support *classical feedback*, where measurement outcomes can influence subsequent quantum operations within a circuit.

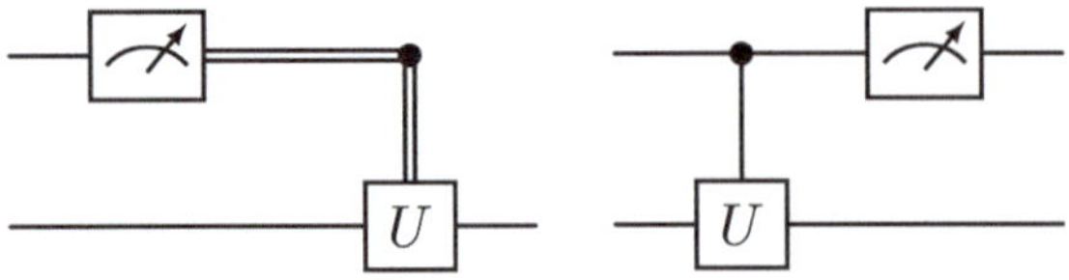

Fig. 10.6 Left: Circuit with gate operation controlled by measurement outcome. Right: Equivalent circuit with measurement deferred to the end of the circuit

Principle of Deferred Measurement

The principle of deferred measurement states that measurements can always be pushed forward or "deferred" to the end of a quantum circuit. If the outcome of a mid-circuit measurement is used to classically control subsequent quantum gates, one may simply replace the gate with controlled quantum gates as shown in Fig. 10.6.

References

1. M. A. Nielsen and I. L. Chuang, *Quantum Computation and Quantum Information: 10th Anniversary Edition*, 1st ed. Cambridge University Press, (2012). https://doi.org/10.1017/CBO9780511976667
2. M. Born, Zur quantenmechanik der stoßvorgänge, Z. Physik, **37**(12), pp. 863–867, (1926). https://doi.org/10.1007/BF01397477
3. J. B. Altepeter, D. F. V. James, P. G. Kwiat, Qubit quantum state tomography. In *Quantum State Estimation*, vol. 649, M. Paris, J. Řeháček, Eds., In Lecture Notes in Physics, **649** , Berlin, Heidelberg: Springer Berlin Heidelberg, pp. 113–145. (2004). https://doi.org/10.1007/978-3-540-44481-7_4.
4. Qiskit contributors, *Qiskit: An Open-source Framework for Quantum Computing*. (2023). https://doi.org/10.5281/ZENODO.2573505
5. V. Bergholm et al., PennyLane: Automatic differentiation of hybrid quantum-classical computations. (2018). *arXiv*. https://doi.org/10.48550/ARXIV.1811.04968

11 Superposition and Entanglement

Superposition is a unique property of quantum computing [1]. A quantum state can exist in a superposition of states. As an example, a single qubit $|\psi\rangle$ can exist in a superposition of $|0\rangle$ and $|1\rangle$. A register of n qubits can exist in an exponentially large superposition of $N = 2^n$ states, which form an orthonormal basis in the Hilbert space $\mathcal{H} : \mathbb{C}^{2^n}$. Representation of an arbitrary quantum state of n qubits requires $O(2^n)$ classical registers. This compact representation as a superposition of basis states enables exponentially efficient representation of data on quantum computers.

The $|0\rangle, |1\rangle$ basis vectors are assigned to the Z basis of qubits by convention. Hardware implementations of qubits will typically allow measurement in only one basis, e.g., the spin state of a fermion. To measure in another basis in hardware implementations, it is convenient to rotate the basis of the qubits and measure them instead of implementing measurement directly in a rotated basis. However, single-qubit rotations do not correspond to arbitrary unitary rotations. Instead, they are limited by the Kronecker product structure of the qubits $\mathcal{H} : \mathbb{C}^{2^n}$.

The Hadamard (H) gate puts qubits in uniform superposition, in which all basis states have the same probability amplitude. As an example, we provide code below to put qubits in uniform superposition and measure their states as shown in the quantum circuit in Fig. 11.1 with the output shown in Fig. 11.2. The Hadamard gate is applied to all three qubits to put them in the uniform superposition state

$$(H \otimes H \otimes H)|000\rangle = \frac{1}{2\sqrt{2}}(|000\rangle + |001\rangle + |010\rangle + \cdots + |111\rangle)$$

All three qubits are then measured, which yields one of the basis states $|000\rangle$,$|001\rangle$,$|010\rangle$, ...,$|111\rangle$. This process is repeated $2^{15} = 32768$ times to obtain the distribution shown in Fig. 11.2. As the number of measurements is increased,

O. M. Raisuddin and S. De, *Quantum Computing for Engineers*,
https://doi.org/10.1007/978-3-032-03325-3_11

the distribution of measurements for qubits in uniform superposition converges to the uniform distribution:

```
#!/usr/bin/python3

from matplotlib import pyplot as plt
import qiskit
from qiskit_aer.primitives import SamplerV2
from qiskit.visualization import plot_histogram

# Create a register of 3 qubits
myQRegister = qiskit.QuantumRegister(3, '\psi')

# Create a register of 3 classical bits
myCRegister = qiskit.ClassicalRegister(3,'ClassicalBits')

# Create a quantum circuit with using myRegister
myCircuit = qiskit.QuantumCircuit(myQRegister, myCRegister)

# Hadamard gates on al qubits
myCircuit.h(myQRegister)

# Measure all the qubits in myQRegister and store state in myCRegis-
ter
myCircuit.measure(myQRegister,myCRegister)

# Simulate the circuit
sampler = SamplerV2()
job = sampler.run([myCircuit],shots=2**15)
result = job.result()[0].data.ClassicalBits.get_counts()

# Plot a bar chart of all the results
plot_histogram(result, title='Uniform Superposition')
plt.show()
```

Entanglement is another property unique to quantum computation [1]. Entangled qubits have a correlated state. A maximal entanglement for a pair of qubits means that the state of one qubit can be fully determined by measuring the state of the other qubit. The simplest set of maximally entangled states are two-qubit states known as the Bell states:

$$\left|\Phi^{+}\right\rangle = \frac{1}{\sqrt{2}}(|00\rangle + |11\rangle), \left|\Phi^{-}\right\rangle = \frac{1}{\sqrt{2}}(|00\rangle - |11\rangle)$$
$$\left|\Psi^{+}\right\rangle = \frac{1}{\sqrt{2}}(|01\rangle + |10\rangle), \left|\Psi^{-}\right\rangle = \frac{1}{\sqrt{2}}(|01\rangle - |10\rangle)$$

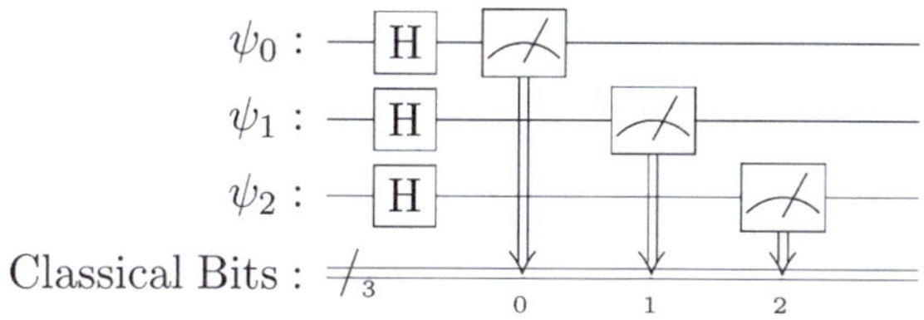

Fig. 11.1 Circuit putting all qubits in superposition

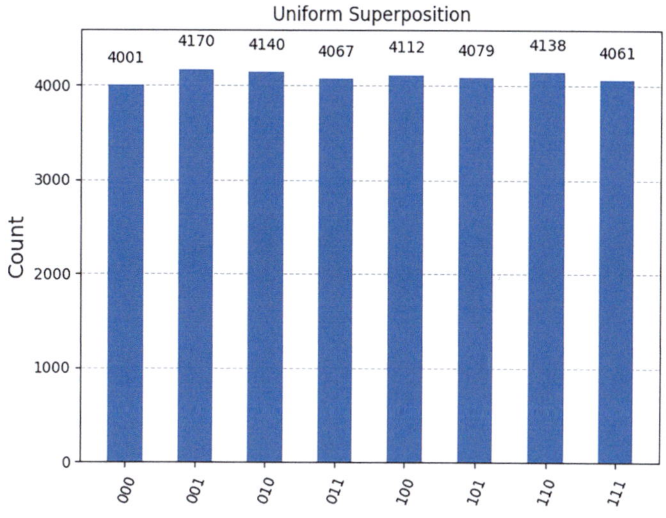

Fig. 11.2 Measurements demonstrating a uniform superposition of basis states

If the first qubit of the state $\left|\Phi^+\right\rangle = \frac{1}{\sqrt{2}}(|01\rangle + |11\rangle)$ is measured as $|0\rangle$, the second qubit is also in the state $|0\rangle$, and if the first qubit is measured as $|1\rangle$, the second qubit is also in the state $|1\rangle$. Note that this is different from the state $|\psi\rangle = |00\rangle + |11\rangle$, which is not an entangled state. The generalization of fully entangled states for more than 2 qubits are known as GHZ states.

Mathematically, entangled qubits cannot be separated into a Kronecker product. The state $|00\rangle$ can be written as $|0\rangle \otimes |0\rangle$: however, it is impossible to separate $\left|\Psi^+\right\rangle$ into such a Kronecker product of Z basis states or any other basis formed by a tensor product of the bases of the individual qubits. Similarly, single-qubit gates cannot rotate an entangled quantum state to a state separable into Kronecker products.

Entangled states are necessary for quantum registers to go from a Hilbert space described by individual $U(2)$ rotations of qubit states to the larger Hilbert space described by $U(2^n)$ rotations of a quantum register.

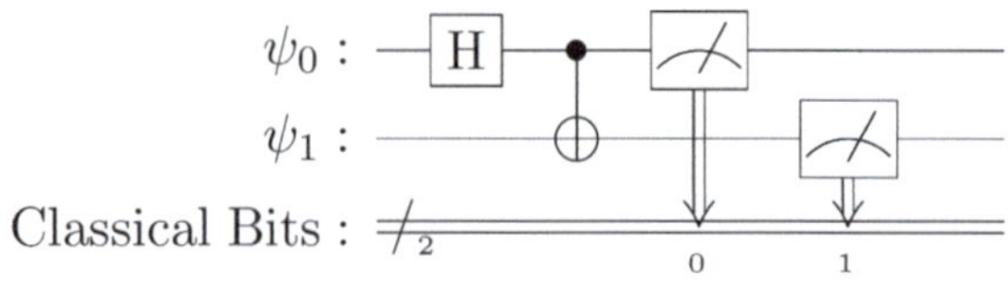

Fig. 11.3 Circuit to entangle two qubits by preparing a Bell state

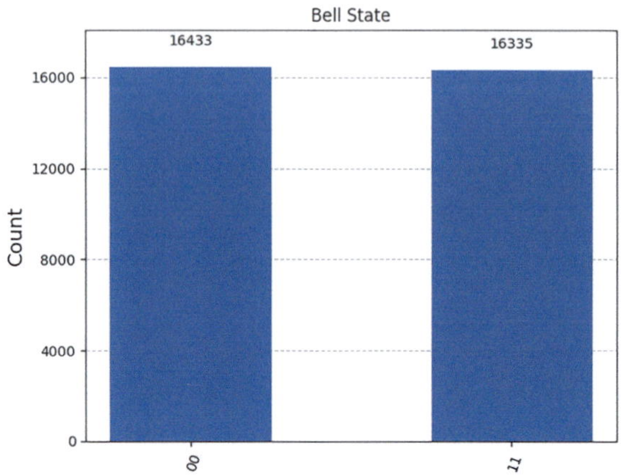

Fig. 11.4 Measurements demonstrating the entanglement of two qubits

In the following code we provide an example of a circuit, shown in Fig. 11.3, creating the entangled state $|\Phi^+\rangle$ with the output measurements shown in Fig. 11.4:

```
#!/usr/bin/python3

from matplotlib import pyplot as plt
import qiskit
from qiskit_aer.primitives import SamplerV2
from qiskit.visualization import plot_histogram

# Create a register of 2 qubits
myQRegister = qiskit.QuantumRegister(2, '\psi')

# Create a register of 2 classical bits
myCRegister = qiskit.ClassicalRegister(2,'ClassicalBits')

# Create a quantum circuit with using myRegister
myCircuit = qiskit.QuantumCircuit(myQRegister, myCRegister)

# Hadamard gates on first qubit
myCircuit.h(0)
```

```
# CNOT gate controlled by first qubit on second qubit
myCircuit.cx(0,1)

# Measure all the qubits in myQRegister and store state in myCRegis-
ter
myCircuit.measure(myQRegister,myCRegister)

# Simulate the circuit
sampler = SamplerV2()
job = sampler.run([myCircuit],shots=2**15)
result = job.result()[0].data.ClassicalBits.get_counts()

# Plot a bar chart of all the results
plot_histogram(result, title='Bell State')
plt.show()
```

Reference

1. M. A. Nielsen, I. L. Chuang, *Quantum Computation and Quantum Information: 10th Anniversary Edition*, 1st ed. Cambridge University Press, (2012). https://doi.org/10.1017/CBO9780511976667

12 Classical and Reversible Computation

All quantum gates (except measurement) are reversible since they are unitary operations. Therefore, the corresponding inverse operation of a gate O is $O^{\dagger}$, where $\dagger$ denotes the Hermitian conjugate of O. For a quantum circuit without a measurement operation, the inverse operation is an application of the Hermitian conjugates of the gates in reverse order.

However, classical computation operations are not necessarily reversible, e.g., the classical logic gates AND and OR. Regardless, it is possible to perform all classical computation operations on quantum computers [1].

Classical Computation on Quantum Computers

First, we note that the classical NAND gate is a universal classical gate, i.e., any classical Boolean operation can be represented as a combination of NAND gates.

We also note that the FANOUT operation may be needed in classical logic circuits involving combinations of NAND gates. Therefore, if the NAND operation and FANOUT operation can be performed on a gate-based quantum computer, then any classical computation can be performed on a quantum computer.

A reversible version of these gates can be formed using ancillary qubits, or "ancilla" qubits. Ancilla qubits are often utilized in quantum computing to implement non-unitary (possibly irreversible) operations as a unitary (reversible) operation. The measured state of ancilla qubits can either be discarded at the end of the computation, used to "post-select" the unmeasured state in the remaining qubits (e.g., continue if ancilla is measured as $|1\rangle$, restart if measured as $|0\rangle$, or used for branching operations for the remainder of a quantum–classical workflow using dynamic circuits.

O. M. Raisuddin and S. De, *Quantum Computing for Engineers*,
https://doi.org/10.1007/978-3-032-03325-3_12

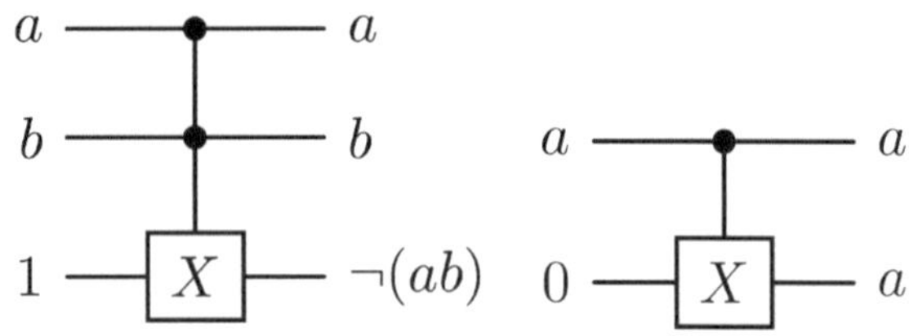

Fig. 12.1 A quantum implementation of Left: a NAND gate; and Right: a FANOUT operation

Even though this approach does not make efficient use of quantum resources, it implies that all classical computation can be performed on quantum computers. Both the NAND and FANOUT operations can be performed using a "controlled-controlled-NOT," or a *ccNOT* gate:

$$ccNOT = \begin{pmatrix} 1 & & & & \\ & \ddots & & & \\ & & 1 & & \\ & & & & 1 \\ & & & 1 & \end{pmatrix}$$

as shown in Fig. 12.1.

Note that measurements are not reversible since they are projection operations, which are rank-deficient (ignoring trivial cases, e.g., I).

Even though this approach does not make efficient use of quantum resources, it implies that all classical computation (using the universal circuit family introduced in Lecture 3: Theory of Computing) can be performed on quantum computers. Both the NAND and FANOUT operations can be performed using a Toffoli gate, or a *ccNOT* gate as shown in Fig. 12.1. Note that although the FANOUT operation copies classical information, this does not generalize to arbitrary quantum states and does not violate the no-cloning theorem. Note also that it is trivial to convert the NAND gate to an AND gate by supplying 0 as an input instead of 1.

Reversible Computing

We have shown above that all classical computing can be performed using quantum circuits, simply using the fact that NAND gates (a universal classical gate set) and FANOUT operations are sufficient to implement any Boolean circuit. However, this strategy requires a large number of ancilla qubits to store intermediate computations, often referred to as "garbage." To alleviate this issue of excess garbage, we introduce here Bennet's "uncompute" trick to remove garbage collected in ancilla qubits.

Consider the task of computing the sequence of AND operations:

$$f(a, b, c) = a \wedge b \wedge c$$

We may compute this function by composing reversible AND gates as shown in Fig. 12.2. For this example $a \wedge b$ may be considered as garbage and $a \wedge b \wedge c$ is the desired result. To make the ancilla containing the garbage $a \wedge b$ available for any subsequent computation, we can "uncompute" $a \wedge b$ and $a \wedge b \wedge c$ as shown in Fig. 12.2.

A general recipe for efficient reversible computation, known as Bennett's uncompute trick [2], is shown in Fig. 12.3 for a general Boolean operation $f : \{0, 1\}^n \to \{0, 1\}^o$ given a corresponding unitary U_f.

This uncompute trick can be used to prove efficient reversible computation of classical computation as follows.

Theorem [2]: A multi-tape Turing machine using time T and space S can be simulated reversibly (or using a quantum circuit) using either.

- $O\left(T^{1+\epsilon}\right)$ time and $O(S \log T)$ space
- $O(T)$ time and $O(ST^{\epsilon})$ space

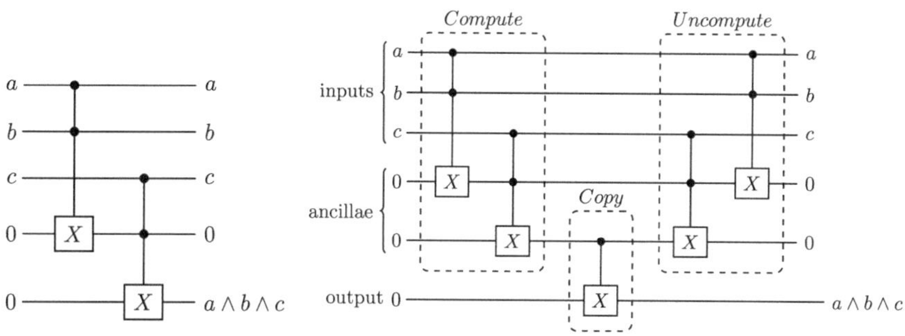

Fig. 12.2 Left: Composing AND gates to compute $a \wedge b \wedge c$; Right: Identical computation with garbage removal

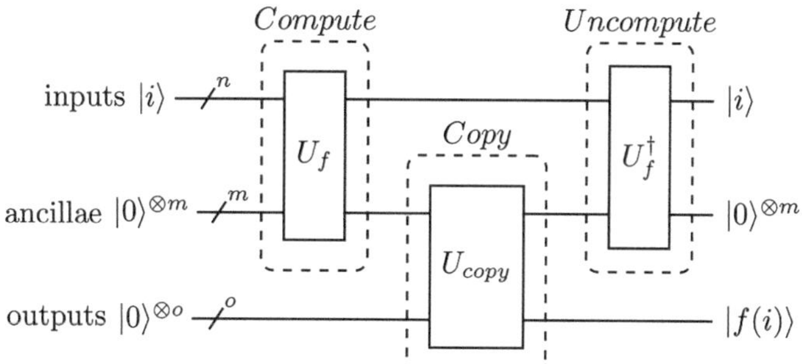

Fig. 12.3 Classical computation performed for n input bits and $o \leq m$ output bits reversibly with ancillae restored to $|0^{\otimes m}\rangle$ for further computation

for any $\epsilon > 0$.

We emphasize that these constructs are motivated by the need to establish relations between classical and quantum computational complexities and do not indicate the usual implementation of quantum algorithms for applied problems.

Quantum Oracles

It is important to note that any arbitrary Boolean function $f : \{0, 1\}^n \rightarrow \{0, 1\}^m$ can be implemented in a reversible manner as shown in Fig. 12.4:

$$U_f |x\rangle |y\rangle = |x\rangle |y \oplus f(x)\rangle$$

Since all the input bitstrings x and y can be inferred from the output, U_f is reversible.

We will refer to an oracle of this form as a quantum "basis state" oracle.

An actual implementation of the gate (or collection of gates) U_f is often unknown and unnecessary for analysis. Oracles are a useful abstraction (and idealization) of some computational procedure. They are used to analyze the computational power of various computing machines and are often used to identify complexity classes. Such examples will be shown in Lecture 15: Simon's Deutsch jozsa, and Bernstein vazirani Algorithms.

To analyze the complexity of some quantum algorithms, it suffices to know that U_f exists with an assumption of "black box" or "oracle" access.

A slightly more useful form of oracles is a quantum "phase oracle." Let's now consider a Boolean function $g : \{0, 1\}^n \rightarrow \{0, 1\}$ that outputs a single bit. A phase oracle simply "marks" the output bits with a phase as shown in Fig. 12.5:

$$U_g |x\rangle = (-1)^{g(x)} |x\rangle$$

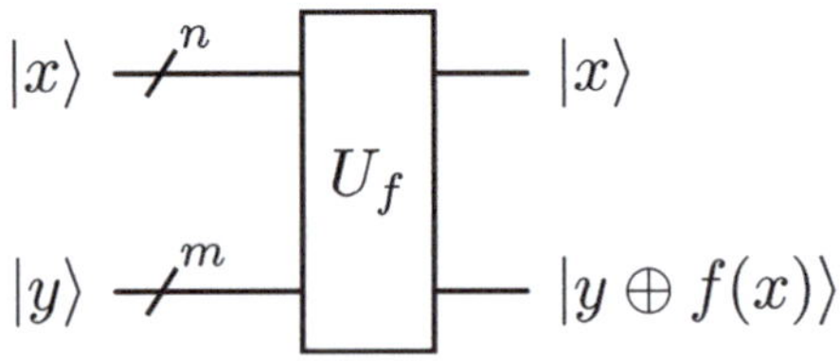

Fig. 12.4 A reversible implementation of a Boolean function $f : \{0, 1\}^n \rightarrow \{0, 1\}^m$ where $x \in \{0, 1\}^n$ and $y, f(x) \in \{0, 1\}^m$ and $\oplus$ denotes a bitwise XOR operation

Fig. 12.5 A quantum phase oracle

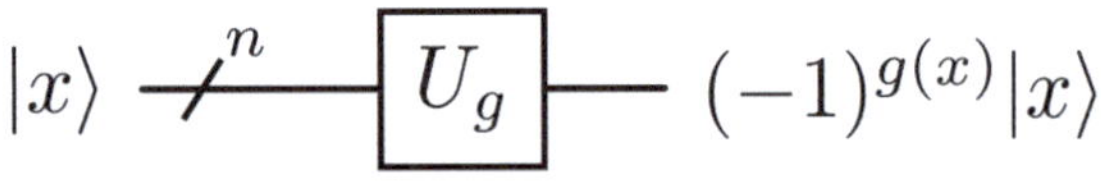

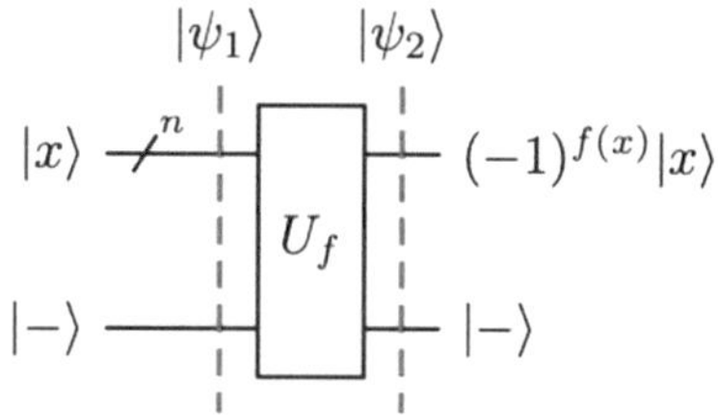

Fig. 12.6 Using a quantum basis state oracle as a phase oracle

We now demonstrate a procedure to convert a quantum basis state oracle into a quantum phase oracle. Consider U_f with $m = 1$ in the following quantum circuit (Fig. 12.6).

Note: $|-\rangle = HX|0\rangle = \frac{1}{\sqrt{2}}(|0\rangle - |1\rangle)$

$$\begin{aligned} |\psi_1\rangle &= |x\rangle \frac{1}{\sqrt{2}}(|0\rangle - |1\rangle) \\ |\psi_2\rangle &= |x\rangle \frac{1}{\sqrt{2}}(|f(x) \oplus 0\rangle - |f(x) \oplus 1\rangle) \\ &= |x\rangle \frac{1}{\sqrt{2}}(|f(x)\rangle - |\neg f(x)\rangle) \end{aligned}$$

Considering each case for $f(x) \in \{0, 1\}$ separately:

$$|\psi_2\rangle = \begin{cases} |x\rangle \frac{1}{\sqrt{2}}(|0\rangle - |1\rangle), f(x) = 0 \\ |x\rangle \frac{1}{\sqrt{2}}(|1\rangle - |0\rangle), f(x) = 1 \end{cases}$$

which simplifies to

$$|\psi_2\rangle = (-1)^{f(x)}|x\rangle|-\rangle$$

which is effectively a phase oracle with an ancilla qubit.

In this chapter, we have provided an overview of connections between classical computing and quantum computing in the context of manipulating classical data in the form of bitstrings. Although this is useful to demonstrate the capability of quantum computers to process classical information and will be used in Lecture 15: Simon's Deutsch Jozsa, and Bernstein Vazirani Algorithms, to show an exponential speedup of quantum computers over classical computers, practical quantum algorithms do not rely on this framework. In the next chapter, we will introduce access models and data representations that are used to develop quantum algorithms for scientific and engineering computation tasks.

References

1. M.A. Nielsen, I.L. Chuang, *Quantum Computation and Quantum Information: 10th Anniversary Edition*, 1st edn (Cambridge University Press, 2012). https://doi.org/10.1017/CBO9780511976667
2. C.H. Bennett, Time/space trade-offs for reversible computation. SIAM J. Comput. **18**(4), 766–776 (1989). https://doi.org/10.1137/0218053

13 Access Models and Data Representation

Classical information is represented as bitstrings in classical computers. The most direct extension to quantum computers—called basis embedding—represents a classical bitstring (e.g., "0101010010") as the quantum state $|0101010010\rangle$, as shown previously. However, this representation is inefficient since it requires the same number of qubits as classical bits and does not exploit the additional quantum degrees of freedom, i.e., the phase and probability amplitudes of the basis states. For example, a vector $a \in \mathbb{R}^{2^n}$ in basis embedding would require $O(2^n)$ qubits, while amplitude encoding can represent it with only n qubits.

Embedding data into the phase of a basis state is known as phase or angle embedding. Similarly, embedding data into the probability amplitude of a basis state is known as amplitude encoding or amplitude embedding [1]. Both amplitude embedding and phase embedding are limited by the definition of quantum states, i.e., the normalization due to the Born rule and the periodicity of phases. Amplitude encoding is the method of choice for quantum algorithms for scientific computing and engineering [2]. Various other encoding methods are used in quantum machine learning applications, and they are discussed in Chap. 38: Quantum Machine Learning In the remainder of this book, we exclusively utilize amplitude embedding unless specified otherwise.

Consider the vector

$$a = \Sigma_{i=0}^{2^n-1} a_i e_i$$

where e_i denotes the ith standard basis vector and n is the number of qubits. If a vector has less than 2^n components, the remainder of the vector may be zero-padded without any loss of generality. An amplitude encoding of this vector is

$$|a\rangle = \frac{1}{\|a\|_2} \sum_{i=0}^{2^n-1} a_i |i\rangle$$

O. M. Raisuddin and S. De, *Quantum Computing for Engineers*,
https://doi.org/10.1007/978-3-032-03325-3_13

where $|i\rangle$ is the ith basis vector of the quantum state. In this notation, $|a\rangle$ represents the quantum state whose amplitudes encode the entries of a, while $|i\rangle$ denotes the ith standard basis vector.

In comparison, a classical register of n bits represents an element of $\{0, 1\}^n$, and a vector $a \in \mathbb{R}^{2^n}$ requires $O(2^n)$ bits for storage, e.g., using the IEEE-754 floating-point format [3]. In amplitude encoding, n qubits suffice, provided the state is normalized.

Access models in quantum computing specify how data is provided to quantum algorithms. The complexity of such algorithms is typically measured by the number of oracle queries—calls to unitary operators that encapsulate data access or operations, abstracting implementation details.

Oracles are essential because they enable algorithms to work with data or subroutines whose explicit implementation may be unknown or even intractable. For example, oracles frequently provide access to matrix entries or prepare quantum states in scientific computing applications. Efficient data access is critical for algorithmic performance and scalability.

Two principal access models for matrices are

- The sparse access model, optimized for matrices with few non-zero entries per row or column.
- The block-encoding model (also known as qubitization), which embeds a matrix as a block of a larger unitary to enable efficient polynomial transformations.

In both access models, oracles serve as unitary operations that act as "black boxes," encapsulating specific operations or data needed by the algorithm. In the case of a linear system problem $Ax = b$ this could be the entries of the matrix A or a method to prepare the vector b.

The complexity of quantum algorithms is often defined by the number of oracle accesses required, as these oracles abstract the implementation details of complex operations, allowing algorithms to access necessary information efficiently. The number of accesses or "queries" to oracles is often referred to as the *query complexity* of a quantum algorithm. A typical underlying assumption is that the information being accessed using the oracle is efficiently computable. For most algorithms, an oracle implementation that scales polylogarithmically in the number of gates with the problem size (or polynomially with the number of qubits) is considered efficient.

There are various oracles for different operations in quantum algorithms. An early access model for matrices, on which the HHL quantum linear system algorithm is based, provides access to an approximation of the unitary matrix operator $U \approx e^{itA}$, where A is a Hermitian matrix, through a Hamiltonian simulation procedure (Chap. 28: Hamiltonian Simulation Techniques). For the ideas and algorithms considered in this book, the more recent and efficient sparse access model [4] and block-encoding model [5, 6] are used to access the entries of a matrix, and a state preparation oracle is used to prepare a quantum state used by an algorithm.

We will first introduce the Sparse Access Model and the Block-encoding model for matrices. Subsequently, we introduce the Hermitian dilation trick to encode an arbitrary matrix as a Hermitian matrix. We then provide a (usually inefficient) recipe to block-encode oracles for matrices as a sum of Pauli strings.

Sparse Access Model

The sparse access model is designed for matrices with a high proportion of zero entries. It uses position and value oracles (unitary operations) to efficiently retrieve non-zero elements:

Position Oracle $\mathcal{O}_A^{pos}$: Provides the position of non-trivially zero elements in a row:

$$\mathcal{O}_A^{pos} : |i, v\rangle \rightarrow |i, j(i, v)\rangle \ \forall\, i, j \in \{1, \ldots, N\}, v \in \{1, \ldots, d\}$$

i : row (or column) number

j : column (or row) number.

v : enumeration of (typically) non-zero entries in row (or column) i.

Value Oracle $\mathcal{O}_A^{val}$: Provides the value of matrix elements:

$$\mathcal{O}_A^{val} : |i, j, z\rangle \rightarrow \left|i, j, A_{i,j} \oplus z\right\rangle \forall z \in \{0, 1\}^{\otimes s}$$

z : initialized bitstring of length s

$A_{i,j} \oplus z \in \{0, 1\}^{\otimes s}$: a bitstring of length s encoding the matrix entry $A_{i,j} \in \mathbb{C}$.

Block-Encoding Model

A more powerful and general access model is the block-encoding model. A block-encoding oracle is access to a unitary (any quantum circuit) of the form

$$U_A = \begin{pmatrix} A/\alpha & * \\ * & * \end{pmatrix}$$

where $\alpha \in \mathbb{R}^+$ is a subnormalization factor, A is the desired matrix to be embedded in U_A, and $*$ are irrelevant entries.

This form is central to methods based on qubitization and quantum signal processing, which are the most powerful and optimal methods for most problems. The main idea is to apply the block-encoded unitary to a state

$$U_A|0\rangle|b\rangle = \begin{pmatrix} A/\alpha & * \\ * & * \end{pmatrix}\begin{pmatrix} b \\ 0 \end{pmatrix} = \begin{pmatrix} Ab/\alpha \\ * \end{pmatrix} = \frac{1}{\alpha}|0\rangle|Ab\rangle + |\perp\rangle$$

Measuring the first qubit in the state $|0\rangle$ indicates a successful matrix–vector multiplication.

Note that although this 2×2 block encoding uses a single ancilla qubit, a register of qubits can also be used for block encoding, which requires m ancilla qubits to be measured as $|0\rangle^{\otimes m}$. We also note that since U_A is unitary, it is necessary that $A/\alpha_2 \le 1$.

The success probability of successfully measuring the ancilla qubits as $|0\rangle^{\otimes m}$ can be defined as follows [2].

Consider any α such that $\|A/\alpha\|_2 \le 1$ so that

$$A/\alpha = \left(0|^{\otimes m} \otimes I_n\right)U_A\left(|0^{\otimes m} \otimes I_n\right)$$

The probability of successfully measuring $|0\rangle^{\otimes m}$ is

$$p\left(|0\rangle^{\otimes m}\right) = \frac{1}{\alpha^2}\|Ab\|^2$$

A block-encoding of this form is often written in short-hand as a "(α, m) block-encoding of A."

A block-encoding may also encode matrices inexactly, i.e.,

$$A - \alpha\left(0^m| \otimes I_n\right)U_A\left(|0^m \otimes I_n\right) \le \epsilon$$

Such block-encodings are often referred to as a "(α, m, ϵ) block encoding of A."

Note that it is always possible to restate a (α, m, ϵ) block-encoding of A as a $(\alpha\beta, m, \epsilon)$ block-encoding of A/β (where $\beta \neq 0$).

Block-encoded oracles also allow addition, products, and tensor products of matrices. Block-encoded matrices may be added using the linear combination of unitaries method, discussed in Chap. 22: Linear Combination of unitaries, and procedures for multiplication are covered in Chap. 26, Matrix Vector Multiplications and Affine Linear Operations. References [5, 11, 12] provide implementation details of these operations. The specifics of oracle implementations for access models are beyond the scope of this book since they are problem-dependent and an active area of research. Recent work has made progress in developing circuits to encode various classes of sparse matrices [7, 13].

We finally note that a sparse access model can be transformed into a block-encoding model using $O(1)$ queries to $\mathcal{O}_A^{pos}$ and $\mathcal{O}_A^{val}$ and $O(\text{poly} \log n)$ additional gates [14].

Hermitian Dilation

We note that various formalisms in quantum computing including oracles for matrices may require the matrices to be Hermitian. In case a matrix A is not Hermitian, its Hermitian dilation H can often be used instead:

$$H = \begin{pmatrix} 0 & A \\ A^{\dagger} & 0 \end{pmatrix}$$

The Hermitian dilation has the same condition number as A, is diagonalizable, and the eigenvalues λ_i of the matrix H are pairwise $\pm\sigma_i$, the singular values σ_i of the matrix A. Therefore, any guarantee of positive- or negative-definiteness of the matrix A is lost. This is apparent by rewriting H as its eigendecomposition using the singular value decomposition (SVD) of A:

$$\begin{pmatrix} 0 & A \\ A^{\dagger} & 0 \end{pmatrix} = \begin{pmatrix} 0 & U\Sigma V^{\dagger} \\ V\Sigma U^{\dagger} & 0 \end{pmatrix} = \frac{1}{2}\begin{pmatrix} U & -U \\ V & V \end{pmatrix}\begin{pmatrix} \Sigma & 0 \\ 0 & -\Sigma \end{pmatrix}\begin{pmatrix} U^{\dagger} & V^{\dagger} \\ -U^{\dagger} & V^{\dagger} \end{pmatrix}$$

Pauli Basis and Decomposition

One may always implement an oracle for a matrix $A \in \mathbb{C}^{2^n \times 2^n}$ by decomposing A in the Pauli basis. First, we note that any Hermitian matrix $A \in \mathbb{C}^{2\times 2}$ may be represented as a linear combination of Pauli matrices, denoted as $\sigma_x = X, \sigma_y = Y, \sigma_z = Z$, and $\sigma_I = I$, where X, Y, Z and I are single-qubit gates as shown in Table 8, as

$$A = \alpha_1\sigma_I + \alpha_2\sigma_x + \alpha_3\sigma_y + \alpha_4\sigma_z$$

where $\alpha_i \in \mathbb{C}$. This can be extended to a general matrix $A \in \mathbb{C}^{2^n}$ as the sum

$$A = \frac{1}{2^n}\sum_{i_1,i_2,\ldots,i_n} \alpha_{i_1,i_2,\ldots,i_n}\sigma_{i_1} \otimes \sigma_{i_2} \otimes \ldots \otimes \sigma_{i_n}$$

where $\alpha_{i_1,i_2,\ldots,i_n} \in \mathbb{C}$ are the coefficients of A in the Pauli basis and $\sigma_{i_j} \in \{I, X, Y, Z\}$ are the identity and Pauli matrices. The coefficients $\alpha_{i_1,i_2,\ldots,i_n}$ can be computed as

$$\alpha_{i_1,i_2,\ldots,i_n} = Tr\big(A \odot (\sigma_{i_1} \otimes \sigma_{i_2} \otimes \ldots \otimes \sigma_{i_n})\big)$$

where $Tr(\cdot)$ is the trace of a matrix, and $\odot$ denotes the Hadamard product (element-wise multiplication) of the matrix entries.

A term of the form $\sigma_{i_1} \otimes \sigma_{i_2} \otimes \ldots \otimes \sigma_{i_n}$ is often referred to as a Pauli string and written as $\sigma_{i_1}\sigma_{i_2}\ldots\sigma_{i_n}$.

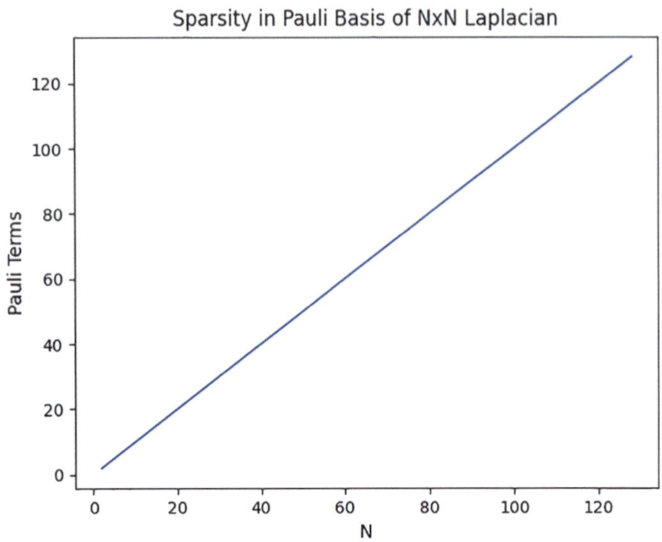

Fig. 13.1 Growth in the number of Pauli terms for a sparse matrix

The linear combination of unitaries (LCU) subroutine, described in Chap. 22, Linear Combination of unitaries, can be used to implement a linear combination of Pauli basis terms [9]. In fact, LCU implementations produce a block-encoded oracle [2, 5]. Although d Pauli basis terms produce at most a d-sparse matrix, i.e., a matrix with at most d non-zero entries in any row or column, a d-sparse matrix in general does not correspond to at most d Pauli basis terms [7]. We provide as an example a code below to decompose the Laplacian matrix into its Pauli basis and show the growth of the number of terms with N in Fig. 13.1. On the contrary, it is known that a Laplacian operator with various boundary conditions can be decomposed into $O(1)$ unitaries with $O\left(n^2\right)$ gates [15] instead of $N = O(2^n)$ gates:

```
#!/usr/bin/python3


from itertools import product
import numpy as np
from qiskit.quantum_info import Operator
from qiskit.circuit import library
import matplotlib.pyplot as plt


def pad_matrix(matrix):
# Pad with 0 to make square matrix
```

```
max_shape = max(matrix.shape[0], matrix.shape [1])
deficiency = int(np.power(2,np.ceil(np.log2(max_shape))) - max_
shape)
if matrix.shape[0] != matrix.shape [1]:
if matrix.shape[0] > matrix.shape [1]:
pad_width = [(0, 0), (0, matrix.shape[0] - matrix.shape [1])]
else:
pad_width = [(0, matrix.shape [1] - matrix.shape[0]), (0, 0)]
matrix = np.pad(matrix, pad_width)
matrix = np.pad(matrix,[(0,deficiency),(0,deficiency)])
return matrix


def decompose_pauli(matrix):
 matrix = pad_matrix(matrix)
 matrix_len = matrix.shape[0]
 nqubits = int(np.log2(matrix_len))


pauli = {
  'x': Operator(library.XGate().to_matrix()),
  'y': Operator(library.YGate().to_matrix()),
  'z': Operator(library.ZGate().to_matrix()),
  'i': Operator(library.IGate().to_matrix())
}


decomposition = {}
for permutation in product(*[list(pauli.keys())]*nqubits):
permutation = "".join(permutation)
base_matrix = pauli[permutation[0]]
for idx in range(1, len(permutation)):
 base_matrix = base_matrix.tensor(pauli[permutation[idx]])
decomposition_component = np.trace(np.dot(base_matrix, matrix)) /
matrix_len
if 0!=decomposition_component:
 decomposition[permutation] = decomposition_component
 return decomposition
max_n = 8
N = [2**n for n in range(1,max_n)]
sparsity = [len(decompose_pauli(2*np.eye(2**n)
          - np.diag(np.ones(2**n-1),-1)
          - np.diag(np.ones(2**n-1),1))) for n in range(1,max_n)]
plt.plot(N,sparsity)
```

```
plt.xlabel('N')
plt.ylabel('Pauli Terms')
plt.title('Sparsity in Pauli Basis of NxN Laplacian')
plt.show()
```

The implementation of efficient oracles for practically relevant problems is an active area of research [7, 13]. An example of a circuit using a sparse matrix access model to encode $U \approx e^{iA}$ where $A = \sum_j A_j$ is a tridiagonal Toeplitz matrix decomposed as a sum of 1-sparse matrices A_j that can be found in [8–10] provide access models for discrete Laplacian matrices.

References

1. M. Weigold, J. Barzen, F. Leymann, M. Salm, Encoding patterns for quantum algorithms. IET Quantum Commun. **2**(4), 141–152 (2021). https://doi.org/10.1049/qtc2.12032
2. L. Lin, Lecture notes on quantum algorithms for scientific computation (2022). Preprint at arXiv 2201.08309. https://doi.org/10.48550/ARXIV.2201.08309
3. IEEE Standard for Floating-Point Arithmetic (2019). https://doi.org/10.1109/IEEESTD.2019.8766229
4. A.W. Harrow, A. Hassidim, S. Lloyd, Quantum algorithm for linear systems of equations. Phys. Rev. Lett. **103**(15), 150502 (2009). https://doi.org/10.1103/PhysRevLett.103.150502
5. S. Chakraborty, A. Gilyén, S. Jeffery, The power of block-encoded matrix powers: improved regression techniques via faster Hamiltonian simulation, in *Leibniz International Proceedings in Informatics*, Schloss Dagstuhl—Leibniz-Zentrum für Informatik, pp. 33:1–33:14 (2019). https://doi.org/10.4230/LIPICS.ICALP.2019.33
6. J.M. Martyn, Z.M. Rossi, A.K. Tan, I.L. Chuang, Grand unification of quantum algorithms. PRX Quantum **2**(4), 040203 (2021). https://doi.org/10.1103/PRXQuantum.2.040203
7. D. Camps, L. Lin, R. Van Beeumen, C. Yang, Explicit quantum circuits for block encodings of certain sparse matrices (2022). Preprint at https://doi.org/10.48550/ARXIV.2203.10236
8. A.C. Vazquez, Quantum algorithma for solving tri-diagonal linear systems of equations. ETH Zurich (2018). http://www.sam.math.ethz.ch/~hiptmair/StudentProjects/CarreraVazquez.Almudena/MScThesis.pdf
9. E. Cappanera, Variational quantum linear solver for finite element problems: a Poisson equation test case. TU Delft (2021). http://resolver.tudelft.nl/uuid:deba389d-f30f-406c-ad7b-babb1b298d87
10. C.J. Trahan, M. Loveland, N. Davis, E. Ellison, A variational quantum linear solver application to discrete finite-element methods. Entropy **25**(4), 580 (2023). https://doi.org/10.3390/e25040580
11. S. Chakraborty, A. Morolia, A. Peduri, Quantum regularized least squares. Quantum **7**, 988 (2023). https://doi.org/10.22331/q-2023-04-27-988
12. A. Gilyén, Y. Su, G.H. Low, N. Wiebe, Quantum singular value transformation and beyond: exponential improvements for quantum matrix arithmetics, in *Proceedings of the 51st Annual ACM SIGACT Symposium on Theory of Computing* (ACM, Phoenix AZ USA, 2019), pp. 193–204. https://doi.org/10.1145/3313276.3316366
13. D. Camps, R. Van Beeumen, Approximate quantum circuit synthesis using block encodings. Phys. Rev. A **102**(5), 052411 (2020). https://doi.org/10.1103/PhysRevA.102.052411
14. D.W. Berry, A.M. Childs, R. Cleve, R. Kothari, R.D. Somma, Exponential improvement in precision for simulating sparse Hamiltonians, in *Proceedings of the Forty-Sixth Annual ACM*

Symposium on Theory of Computing, in STOC '14 (Association for Computing Machinery, New York, NY, USA, 2014), pp. 283–292. https://doi.org/10.1145/2591796.2591854

15. O.M. Raisuddin, S. De, qRLS: quantum relaxation for linear systems in finite element analysis. Eng. Comput. (2024). https://doi.org/10.1007/s00366-024-01975-3

14 Limitations of Quantum Computers

Quantum physics places some fundamental limits on possible operations using quantum computing. A well-known limitation is the no-cloning theorem [1], which states that an arbitrary quantum state cannot be used to make an exact, independent copy of itself. More precisely, the map $U|\phi\rangle|\psi\rangle \rightarrow e^{i\alpha}|\psi\rangle|\psi\rangle$ is not possible in general for arbitrary $|\psi\rangle \in \mathbb{C}^n$, $|\phi\rangle \in \mathbb{C}^n$ where $U \in \mathbb{C}^{2n}$.

The no-deletion theorem complements the no-cloning theorem, which prohibits information in a quantum state from being erased (using unitary operations). Note that imperfect copies are still possible to create, with known bounds on the error [2–4], or perfect copies can be made if the quantum state is fully known.

The normalization of a state and the periodicity of phase can also be considered limitations on the state. All gate operations are unitary and linear. To apply non-unitary and nonlinear operations, a projection onto a subspace of the overall linear space must be considered. This is the central idea behind block-encoding. This property is exploited in various quantum computing algorithms, at the expense of ancilla qubits and a non-zero probability of failure [5].

Amplifying the probability of a desirable quantum state among a superposition of undesirable states can also pose a challenging limitation. If the amplitude of the desired state is exponentially small, the probability of obtaining the desired state cannot be boosted without an exponential overhead [6]. This is known as the post-selection problem [7].

Getting data in and out of a quantum register is also a challenging problem. I/O is expensive on quantum computers: preparing or reading out an arbitrary quantum state scales as $O(2^n)$. Furthermore, reading out a quantum state with the phases requires quantum state tomography [8]. The quantum version of random-access memory, QRAM, has been proposed to access and store classical data on quantum computers efficiently and is an active area of research [9].

O. M. Raisuddin and S. De, *Quantum Computing for Engineers*,
https://doi.org/10.1007/978-3-032-03325-3_14

Some problems have been proven to exhibit no quantum speedup compared to classical computing. A well-known result is the quantum no-fast-forwarding theorem [10] which states that the optimal scaling for an arbitrary Hamiltonian simulation for time T is $O(T)$, in the worst case. The standard proxy for this proof is the parity problem: given oracle access to a string of N bits, computing the parity classically requires N queries, while the optimal quantum algorithm achieves no better than $N/2$ queries [11]. Thus, no exponential or even superpolynomial quantum advantage is possible for this task. However, for specific classes of Hamiltonians—such as certain positive-definite or structured systems, sublinear time evolution via fast-forwarding is possible [5].

Computing expectation values of observables is a fundamental subroutine in many quantum algorithms. Given a quantum state $|\psi\rangle$ and an observable O, the goal is to estimate $\mathrm{O}_{\psi} = \langle\psi|\mathrm{O}|\psi\rangle$ to within additive precision ϵ. In general, the number of measurements required to achieve standard deviation ϵ scales $O\big(1/\epsilon^2\big)$, due to statistical sampling. However, using quantum phase estimation or related techniques, this can be improved to the so-called Heisenberg limit, where the cost scales as $O(1/\epsilon)$.

Despite these limitations, there is potential for quantum computing to make an impact on scientific computation and engineering problems. Classical computers and algorithms are simply not capable of storing and processing large quantum simulations, making quantum computers the only viable option for general quantum simulation problems [12] despite the no-fast-forwarding theorem. Algorithms typically need to be modified to be amenable to quantum computing. For example, [13] uses a Carleman linearization of a nonlinear ordinary differential equation to obtain an exponential speedup in the number of unknowns, and [14] presents a novel data encoding scheme for the efficient implementation of a quantum lattice Boltzmann method.

There exist many other no-go theorems for quantum computing. It is important to be mindful of these results when developing algorithms to remain within the confines of what is physically computable, since any computer we can build must follow the laws of physics.

References

1. W.K. Wootters, W.H. Zurek, A single quantum cannot be cloned. Nature **299**(5886), 802–803 (1982). https://doi.org/10.1038/299802a0
2. V. Bužek, M. Hillery, Quantum copying: beyond the no-cloning theorem. Phys. Rev. A **54**(3), 1844–1852 (1996). https://doi.org/10.1103/PhysRevA.54.1844
3. V. Bužek, M. Hillery, Universal optimal cloning of arbitrary quantum states: from qubits to quantum registers. Phys. Rev. Lett. **81**(22), 5003–5006 (1998). https://doi.org/10.1103/PhysRevLett.81.5003
4. V. Buzek, M. Hillery, Quantum cloning. Phys. World **14**(11), 25–30 (2001). https://doi.org/10.1088/2058-7058/14/11/28
5. D. An, J.P. Liu, D. Wang, Q. Zhao, A theory of quantum differential equation solvers: limitations and fast-forwarding (2022). Preprint at arXiv https://doi.org/10.48550/ARXIV.2211.05246.

6. G. Brassard, P. Høyer, M. Mosca, A. Tapp, Quantum amplitude amplification and estimation, in *Contemporary Mathematics*, vol. 305, eds. by S.J. Lomonaco, H.E. Brandt (Providence, Rhode Island: American Mathematical Society, 2002), pp. 53–74. https://doi.org/10.1090/conm/305/05215
7. S. Aaronson, Quantum computing, postselection, and probabilistic polynomial-time (2004). Preprint at arXiv arXiv:quant-ph/0412187
8. M. Cramer et al., Efficient quantum state tomography. Nat. Commun. **1**(1), 149 (2010). https://doi.org/10.1038/ncomms1147
9. K. Phalak, A. Chatterjee, S. Ghosh, Quantum random access memory for dummies (2023). Preprint at arXiv https://doi.org/10.48550/ARXIV.2305.01178.
10. D.W. Berry, G. Ahokas, R. Cleve, B.C. Sanders, Efficient quantum algorithms for simulating sparse Hamiltonians. Commun. Math. Phys. **270**(2), 359–371 (2007). https://doi.org/10.1007/s00220-006-0150-x
11. E. Farhi, J. Goldstone, S. Gutmann, M. Sipser, Limit on the speed of quantum computation in determining parity. Phys. Rev. Lett. **81**(24), 5442–5444 (1998). https://doi.org/10.1103/PhysRevLett.81.5442
12. R.P. Feynman, Quantum mechanical computers. Found. Phys. **16**(6), 507–531 (1986). https://doi.org/10.1007/BF01886518
13. J.P. Liu, H.Ø. Kolden, H.K. Krovi, N.F. Loureiro, K. Trivisa, A.M. Childs, Efficient quantum algorithm for dissipative nonlinear differential equations. Proc. Natl. Acad. Sci. U.S.A. **118**(35), e2026805118 (2021). https://doi.org/10.1073/pnas.2026805118
14. M.A. Schalkers, M. Möller, On the importance of data encoding in quantum Boltzmann methods (2023). Preprint at arXiv arXiv:2302.05305. Accessed 07 Nov 2023

Simon's, Deutsch–Jozsa, and Bernstein–Vazirani Algorithms

15

In this chapter, we introduce three abstract computational problems to demonstrate a clear advantage of quantum computers over classical computers in a black-box setting.

Although these algorithms do not have any known applied utility, they distinguish the complexity classes of quantum computers. At the end of this chapter, we will touch upon the Hidden subgroup problem, demonstrating how it is a stencil for exponential quantum speedups with an example.

In this chapter, we will make extensive use of the following identity to represent Hadamard gates on n qubits:

$$H^{\otimes n} = \sum_{x \in \{0,1\}^n} |\psi_x\rangle\langle x|$$

$$|\psi_x\rangle = \frac{1}{\sqrt{2^n}} \sum_{x \in \{0,1\}^n} (-1)^{x \cdot y} |y\rangle$$

where $x \cdot y$ is a binary inner product $x \cdot y = x_0 y_0 + x_1 y_1 + \ldots + x_{n-1} y_{n-1}$.

Applying this identity to a basis state $|i\rangle$ yields

$$H^{\otimes n}|i\rangle = 2^{-\frac{n}{2}} \sum_{j=0}^{2^n-1} (-1)^{i \cdot j} |j\rangle$$

Deutsch–Jozsa Algorithm

The Deutsch–Jozsa algorithm [1] demonstrates an exponential speedup over classical computing. The computational problem to be solved is defined as follows:

O. M. Raisuddin and S. De, *Quantum Computing for Engineers*,
https://doi.org/10.1007/978-3-032-03325-3_15

Given a black-box procedure (a.k.a. oracle) that implements a function mapping, a binary string of length n to a Boolean value

$$f : \{0, 1\}^n \rightarrow \{0, 1\}$$

s.t. f *is known to be constant:*

$$f(b) = 1 \; or \; f(b) = 0 \; \forall \; b \in \{0, 1\}^n$$

or balanced:

$$f(b) = \begin{cases} 0 \; \forall \; b_1 \\ 1 \; \forall \; b_2 \end{cases}$$

where $b_1 \cap b_2 = \emptyset$, $b_1 \cup b_2 = \{0, 1\}^n$, *and* $|b_1| = |b_2| = 2^{n-1}$.

Determine whether f *is constant or balanced.*

On a classical computer, this algorithm requires $2^{n-1}+1$ evaluations of f (oracle queries) in the worst case.

A quantum oracle for f requires it to be reversible. As explained in Chap. 12, Classical and Reversible Computation, we can assume access to an oracle of the form shown in Fig. 15.1.

To solve this problem using a quantum computer, we take advantage of quantum parallelism by preparing a superposition of inputs. The problem can be solved by executing the following circuit in Fig. 15.2.

There are several methods to derive the proof of this algorithm. In the proof that follows, we use an orthogonality argument. Tracing the steps of this circuit:

$$|\psi_1\rangle = \left(I^{\otimes n} \otimes (HX)\right)|0\rangle^{\otimes n}|0\rangle$$

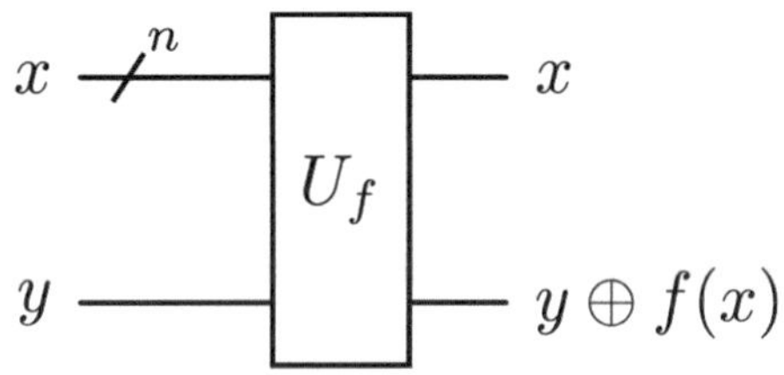

Fig. 15.1 Quantum oracle for f

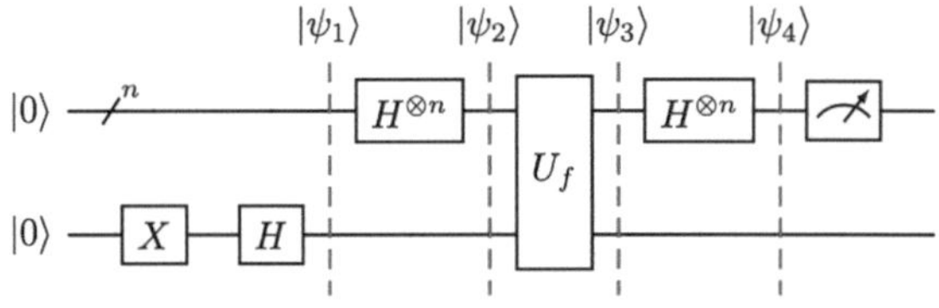

Fig. 15.2 Quantum circuit for the Deutsch–Jozsa algorithm

$$= \frac{1}{\sqrt{2}}|0\rangle^{\otimes n}(|0\rangle - |1\rangle) = |0\rangle^{\otimes n}|0\rangle$$

Note that we have effectively transformed U_f into a phase oracle, i.e.,

$$U_f|x\rangle|-\rangle = (-1)^{f(x)}|x\rangle|-\rangle$$

We will now evaluate f for all possible inputs $x \in \{0, 1\}^n$ using a superposition state (using the identity for Hadamard gates provided earlier):

$$|\psi_2\rangle = \left(H^{\otimes n} \otimes I\right)|\psi_1\rangle = 2^{-\frac{n}{2}} \sum_{i=0}^{2^n-1} |i\rangle|-\rangle$$

The quantum oracle U_f is now applied to this superposition of all possible inputs

$$|\psi_3\rangle = 2^{-\frac{n}{2}} \sum_{i=0}^{2^n-1} (-1)^{f(i)}|i\rangle|-\rangle$$

Now consider the two cases when f is constant:

$$f(i) = \begin{cases} 1 \therefore (-1)^{f(i)} = -1 \\ 0 \therefore \ (-1)^{f(i)} = 1 \end{cases} \quad \forall\, i$$

Therefore, for the constant case, we can equivalently write

$$|\psi_3\rangle = (-1)^{f(0)}2^{-\frac{n}{2}} \sum_{i=0}^{2^n-1} |i\rangle|-\rangle$$

which is simply an overall phase. Proceeding further with the constant case, applying Hadamard gates to the first register now yields

$$|\psi_4\rangle = (-1)^{f(0)}|0\rangle^{\otimes n}|-\rangle$$

Therefore, for the constant case, all the bits in the first register must equal zero. To complete the proof, we now need to show that this does not hold for the balanced case.

Now let's return to the state $|\psi_3\rangle$

$$|\psi_3\rangle = 2^{-\frac{n}{2}} \sum_{i=0}^{2^n-1} (-1)^{f(i)}|i\rangle|-\rangle$$

Let's measure the overlap of this state between the constant and balanced cases, i.e.,

$$\left\langle \psi_3^{constant} | \psi_3^{balanced} \right\rangle = \left[(-1)^{f(0)}2^{-\frac{n}{2}} \sum_{i=0}^{2^n-1} \langle i|\langle -| \right] \left[2^{-\frac{n}{2}} \sum_{j=0}^{2^n-1} (-1)^{f(j)}|j\rangle|-\rangle \right]$$

$$
\begin{aligned}
&= (-1)^{f(0)} 2^{-n} \left[\sum_{i=0}^{2^n-1} \langle i| \right] \left[\sum_{j=0}^{2^n-1} (-1)^{f(j)} |j\rangle \right] \\
&= (-1)^{f(0)} 2^{-n} \sum_{i=0}^{2^n-1} \sum_{j=0}^{2^n-1} (-1)^{f(j)} \langle i|j\rangle \\
&= (-1)^{f(0)} 2^{-n} \sum_{j=0}^{2^n-1} (-1)^{f(j)} \langle j|j\rangle \\
&= (-1)^{f(0)} 2^{-n} \sum_{j=0}^{2^n-1} (-1)^{f(j)}
\end{aligned}
$$

Since $f(j)$ is balanced, $\sum_{j=0}^{2^n-1} (-1)^{f(j)} = 0$. Therefore,

$$
\left\langle \psi_3^{constant} | \psi_3^{balanced} \right\rangle = 0
$$

implying that $\left|\psi_3^{constant}\right\rangle \perp \left|\psi_3^{balanced}\right\rangle$.

Since $\left|\psi_3^{constant}\right\rangle$ is orthogonal to $\left|\psi_3^{balanced}\right\rangle$, they will remain orthogonal if any unitary transformation is applied to both states, i.e.,

$$
U\left|\psi_3^{constant}\right\rangle \perp U\left|\psi_3^{balanced}\right\rangle \forall\, U \in U(2^n)
$$

Since the transformation from $|\psi_3\rangle \to |\psi_4\rangle$ is unitary, i.e., $H^{\otimes n} \otimes I$,

$$
\left|\psi_4^{constant}\right\rangle \perp \left|\psi_4^{balanced}\right\rangle
$$

Since we have already established that the state $\left|\psi_4^{constant}\right\rangle = |0\rangle^{\otimes n}|-\rangle$ up to an overall phase:

$$
|0\rangle^{\otimes n} \perp \left|\psi_4^{balanced}\right\rangle
$$

therefore, for the balanced case, a measurement of the first register cannot yield all zeros.

To summarize, executing the circuit shown in Fig. 15.1 and measuring the first register solves the Deutsch–Jozsa problem. Measuring any qubit in a non-zero state indicates that the function is balanced, and measuring all qubits in the first register in the zero state indicates that the function is constant.

Since this procedure requires only one query to an oracle, this demonstrates an exponential improvement in query complexity over any classical method for this problem.

Bernstein–Vazirani Problem

The Bernstein–Vazirani algorithm [2] is designed to discover a string hidden in a function. The Bernstein–Vazirani algorithm demonstrates a linear (polynomial) speedup over classical query complexity.

Problem statement:

Given $f : \{0,1\}^n \rightarrow \{0,1\}$ *and a secret string* $s \in \{0,1\}^n$, $f(x) = parity(AND(x,s))$. *Determine s.*

$f(x)$ can alternatively be defined as a bitwise dot product between x and s modulo 2:

$$f(x) = x \cdot s = x_0 s_0 \oplus x_1 s_1 \oplus \ldots \oplus x_{n-1} s_{n-1} = \mathrm{mod}_2 \sum_{i=0}^{n-1} x_i s_i$$

s can be determined classically one bit at a time by querying f using the following strings:

$$s_i = f\left(\mathrm{bin}\left(2^i\right)\right) \ \forall\, i \in [0, n-1]$$

which requires n queries to a classical oracle of f to determine all bits of s.

Given access to f as a quantum oracle U_f, the string s may be determined using one query to U_f, indicating a linear speedup over classical query complexity.

To determine s using a quantum procedure similar to the Deutsch–Jozsa problem, we assume access to a quantum oracle of the form:

$$U_f|x\rangle|b\rangle = |x\rangle|b \oplus f(x)\rangle.$$

and using the same technique employed for the Deutsch–Jozsa algorithm, we transform this oracle into a phase oracle by choosing $|b\rangle = |-\rangle$:

$$U_f|x\rangle|-\rangle = (-1)^{f(x)}|x\rangle|-\rangle$$

The circuit for solving this problem is similar to that for the Deutsch–Jozsa problem and is shown in Fig. 15.3.

Tracing this algorithm step-by-step, we get

$$|\psi_1\rangle = \left(I^{\otimes n} \otimes HX\right)|0\rangle^{\otimes n}|0\rangle = |0\rangle^{\otimes n}|-\rangle$$

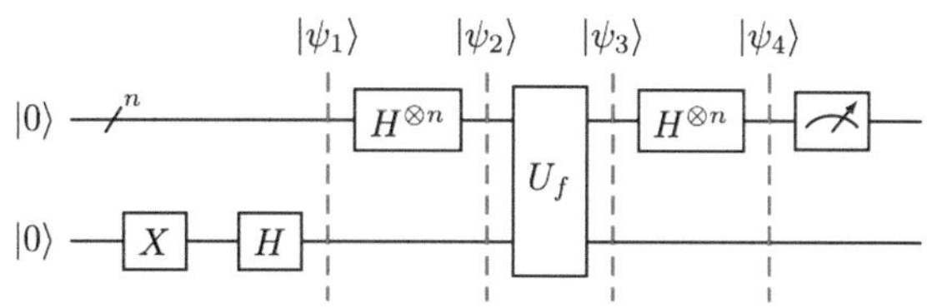

Fig. 15.3 Quantum circuit to solve the Bernstein–Vazirani problem

$$|\psi_2\rangle = 2^{-\frac{n}{2}} \sum_{i=0}^{2^n-1} |i\rangle|-\rangle$$

$$|\psi_3\rangle = 2^{-\frac{n}{2}} \sum_{i=0}^{2^n-1} (-1)^{f(i)} |i\rangle|-\rangle$$

By definition of $f(i)$ we get

$$|\psi_3\rangle = 2^{-\frac{n}{2}} \sum_{i=0}^{2^n-1} (-1)^{i\cdot s} |i\rangle|-\rangle$$

In the next step we use the identity:

$$H^{\otimes n}|i\rangle = 2^{-\frac{n}{2}} \sum_{j=0}^{2^n-1} (-1)^{i\cdot j} |j\rangle$$

where $|i\rangle, |j\rangle \in \mathbb{C}^{2^n}$ are standard basis states.

Applying the final Hadamard gates, we get

$$\begin{aligned}
|\psi_4\rangle &= 2^{-\frac{n}{2}} \sum_{i=0}^{2^n-1} (-1)^{i\cdot s} \left(H^{\otimes n}|i\rangle\right)|-\rangle \\
&= 2^{-\frac{n}{2}} \sum_{i=0}^{2^n-1} (-1)^{i\cdot s} \left(2^{-\frac{n}{2}} \sum_{j=0}^{2^n-1} (-1)^{i\cdot j} |j\rangle\right)|-\rangle \\
&= 2^{-n} \sum_{i=0}^{2^n-1} \sum_{j=0}^{2^n-1} (-1)^{i\cdot s}(-1)^{i\cdot j} |j\rangle|-\rangle
\end{aligned}$$

We now use the following identities:

$$(-1)^{i\cdot s}(-1)^{i\cdot j} = (-1)^{(i\cdot s)\oplus(i\cdot j)} = (-1)^{i\cdot(s\oplus j)}$$

where $\oplus$ indicates a bitwise XOR operation, to arrive at

$$|\psi_4\rangle = 2^{-n} \sum_{i=0}^{2^n-1} \sum_{j=0}^{2^n-1} (-1)^{i\cdot(s\oplus j)} |j\rangle|-\rangle$$

Now we can compute the probability amplitude of the state $|s\rangle|-\rangle$ by choosing $j = s$. First, we note that $s \oplus s = 0$. Furthermore, $i \cdot 0 = 0$. Therefore, we can now simplify

$$\begin{aligned}
|\psi_4\rangle &= 2^{-n} \sum_{i=0}^{2^n-1} (-1)^0 |s\rangle|-\rangle \\
|\psi_4\rangle &= 2^{-n} \sum_{i=0}^{2^n-1} |s\rangle|-\rangle \\
|\psi_4\rangle &= |s\rangle|-\rangle
\end{aligned}$$

and see that the probability amplitude of $|s\rangle$ is 1. Therefore, using a completeness argument, we know that all other states must have a probability amplitude of 0. We can conclude that measuring the first register will directly yield the string s.

Simon's Problem

Simon's problem [3] is similar to the Bernstein–Vazirani problem; however, it demonstrates an exponential speedup over classical computing. There are several variants of Simon's problem, stated as follows. We will attempt to combine all these (equivalent variants).

We provide two equivalent statements of the problem as Variant 1a and Variant 2a. If the problem asks for an additional output, we provide Variant 1b and Variant 2b as equivalent extensions of the problem.

Variant 1a (decision problem):

Given $f : \{0,1\}^n \to \{0,1\}^n$ *and* $s, x, y \in \{0,1\}^n$ *s.t.* $\forall\, x, y$ $f(x) = f(y)$ *iff* $y = x \oplus s$.

Determine whether $s = 0^n$ *or* $s \neq 0^n$.

Variant 1b:

Given $s \neq 0^n$, *find* s.

Variant 2a (decision problem):

Given $f : \{0,1\}^n \to \{0,1\}^n$ *and* $s, x, y \in \{0,1\}^n$ *s.t.* $\forall\, x, y$ $f(x) = f(y)$ *iff* $y = x \oplus s$.

Determine whether f *is a* $1-1$ *or* $2-1$ *function.*

Variant 2b:

If f *is a* $2-1$ *function, determine the secret string* s.

One deterministic classical algorithm to solve Simon's problem is to query $f(x)$ for at most 2^{n-1} distinct values of $x \in \{0,1\}^n$. If there is a "collision," i.e., $f(x) = f(y)$, it will necessarily be apparent by comparing all the computed strings $f(x)$, solving Variants 1a and 2a of the problem.

Variants 1b and 2b can also be solved by simply using the colliding pair of inputs x, y and computing $x \oplus y = x \oplus x \oplus s = 0^n \oplus s = s$. Therefore, the worst-case scenario of this algorithm is $2^{n-1} + 1$ queries of $f(x)$ to solve all variants of the problem.

This problem may also be solved using a probabilistic classical algorithm utilizing the "Birthday Paradox." The lower bound for a randomized classical algorithm requires $O\left(2^{\frac{n}{2}}\right)$ queries of $f(x)$. In either case, the classical query complexity is known to scale exponentially with the length of the string.

If $f(x)$ is provided as a quantum oracle U_f of the form

$$U_f|x\rangle|0\rangle^{\otimes n} = |x\rangle|f(x)\rangle$$

this problem may be solved with $O(n)$ queries to U_f and $O\left(n^3\right)$ classical post-processing steps, indicating an exponential speedup. The quantum circuit for solving this problem is shown in Fig. 15.4.

Let's trace the quantum states of this algorithm as labeled in Fig. 15.4. Using the identity for the Hadamard gate

$$|\psi_1\rangle = (H \otimes I)|0\rangle^{\otimes n}|0\rangle^{\otimes n} = 2^{-\frac{n}{2}} \sum_{i=0}^{2^n-1} |i\rangle|0\rangle^{\otimes n}$$

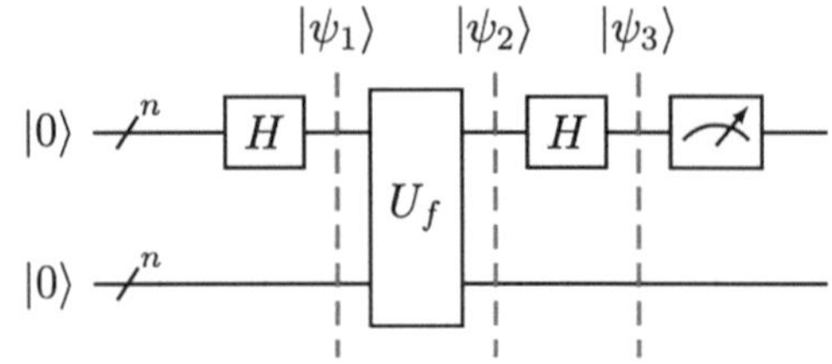

Fig. 15.4 Quantum circuit for solving Simon's problem

Using the definition of $U_f U_f$

$$|\psi_2\rangle = U_f 2^{-\frac{n}{2}} \sum_{i=0}^{2^n-1} |i\rangle|0\rangle^{\otimes n} = 2^{-\frac{n}{2}} \sum_{i=0}^{2^n-1} |i\rangle|f(i)\rangle$$

At this point, let's analyze $|\psi_2\rangle$ before proceeding further. Consider a measurement of the second register. Measuring a bitstring $f(j)$ in the second register implies that the overall system can be in either of the following two states:

$$\begin{array}{ll} \frac{1}{\sqrt{2}}(|j\rangle + |j \oplus s\rangle)|f(j)\rangle & \textit{if } s \neq 0^n \\ |j\rangle|f(j)\rangle & \textit{if } s = 0^n \end{array}$$

Let's proceed with the case $s \neq 0^n$. We may now rewrite $|\psi_2\rangle$ as

$$|\psi_2\rangle = 2^{-\frac{n+1}{2}} \sum_{i=0}^{2^n-1} (|i\rangle + |i \oplus s\rangle)|f(i)\rangle$$

We now apply the final Hadamard gates and use the identity

$$H^{\otimes n}|i\rangle = 2^{-\frac{n}{2}} \sum_{j=0}^{2^n-1} (-1)^{i \cdot j}|j\rangle$$

We can now simply drop the second register as a "don't care" term and apply the Hadamard gates to obtain

$$\begin{aligned} &(H^{\otimes n} \otimes I)2^{-\frac{n+1}{2}} \sum_{i=0}^{2^n-1} (|i\rangle + |i \oplus s\rangle)|f(i)\rangle \\ &= 2^{-\frac{n+1}{2}} \sum_{i=0}^{2^n-1} \left(2^{-\frac{n}{2}} \sum_{j=0}^{2^n-1} (-1)^{i \cdot j}|j\rangle + 2^{-\frac{n}{2}} \sum_{j=0}^{2^n-1} (-1)^{(i \oplus s) \cdot j}|j\rangle\right)|f(i)\rangle \\ &= 2^{-\frac{n+1}{2}} \sum_{i=0}^{2^n-1} \left(2^{-\frac{n}{2}} \sum_{j=0}^{2^n-1} (-1)^{i \cdot j}|j\rangle + 2^{-\frac{n}{2}} \sum_{j=0}^{2^n-1} (-1)^{i \cdot j}(-1)^{s \cdot j}|j\rangle\right)|f(i)\rangle \\ &= 2^{-\frac{2n+1}{2}} \sum_{i=0}^{2^n-1} \sum_{j=0}^{2^n-1} (-1)^{i \cdot j}\left(1 + (-1)^{s \cdot j}\right)|j\rangle|f(i)\rangle \end{aligned}$$

Noting that $1 + (-1)^{s \cdot j} \neq 0$ only if $s \cdot j = 0$, measuring the first register yields $|j\rangle$ s.t. $s \cdot j = 0$. This does not directly reveal s. However, by sampling $n-1$ linearly independent samples $\overset{0}{j}, \overset{1}{j}, \ldots \overset{n-2}{j}$ we can construct a homogeneous linear system problem of the form

$$\begin{pmatrix} \overset{0}{j}_0 & \overset{0}{j}_1 & \cdots & \overset{0}{j}_{n-1} \\ \overset{1}{j}_0 & \overset{1}{j}_1 & \cdots & \overset{1}{j}_{n-1} \\ \vdots & \vdots & \ddots & \vdots \\ \overset{n-2}{j}_0 & \overset{n-2}{j}_1 & \cdots & \overset{n-2}{j}_{n-1} \end{pmatrix} \begin{pmatrix} s_0 \\ s_1 \\ \vdots \\ s_{n-1} \end{pmatrix} = \begin{pmatrix} 0 \\ 0 \\ \vdots \\ 0 \end{pmatrix}$$

The number of bitstrings that need to be sampled to obtain such linearly independent bitstrings is $O(n)$. This linear system can be solved for a non-trivial solution using $O(n^3)$ floating-point operations on a classical computer to reveal the string s.

Therefore, the overall complexity of Simon's algorithm is $O(n)$ quantum oracle access and $O(n^3)$ classical computing operations, leading to a polynomial time solution. This demonstrates an exponential improvement in query complexity over classical methods for solving Simon's problem.

Hidden Subgroup Problem

We note that Simon's problem is an instance of a very particular type of problem known as an Abelian Hidden Subgroup Problem (HSP). We define the Abelian HSP below.

Consider a finite group $G : G \times G \to G$ and a subgroup $H \subset G$, i.e., $H : H \times H \to H$ with a group operation $\odot$. The group operation is Abelian (commutative) if $g_i \odot g_j = g_j \odot g_i \ \forall \ g_i, g_j \in G$.

Consider also a function on G as $f : G \to A$, where A is a finite set. f is said to *hide* a subgroup H if

$f(g_i) = f(g_j)$ *iff* $g_i \odot H = g_j \odot H$.

Abelian Hidden Subgroup Problem: *Given oracle access to f and the members of a group G, find the subgroup H that is hidden by an Abelian group operation $\odot$.*

We can now draw parallels to see how the problem is an instance of the HSP. The group G is formed by bitstrings $\{0, 1\}^n$. The group operation $\oplus$ is Abelian. The finite set A also happens to be $\{0, 1\}^n$. The hidden subgroup H is $\{0^{\otimes n}, s\}$:

$$g_i \oplus H = \{g_i \oplus 0^{\otimes n}, g_i \oplus s\} = \{g_i, g_i \oplus s\} = \{g_i, s \oplus g_i\}$$

Evaluating f on this left coset or right coset $\{g_i, g_i \oplus s\} = \{g_i, s \oplus g_i\}$ yields

$$f(g_i) = f(g_i \oplus s)$$

which is the definition of f, i.e., f does indeed hide H by definition.

Quantum computers are known to have an exponential speedup (polynomial time solution) over classical problems for the Abelian HSP. Whether the non-Abelian version of this problem exhibits a similar speedup is an open problem. A prominent application is Shor's algorithm [4], which reduces integer factorization and discrete logarithm to instances of the Abelian HSP over cyclic groups. This approach yields an exponential quantum speedup for these problems compared to all known classical algorithms.

The three problems outlined in this chapter demonstrate the fundamentally different nature of quantum computing in contrast to classical computing. Although these problems themselves do not solve any applied problem in computing, they foreshadow the potential speedups that can be enabled by quantum computing and provide a separation of complexity classes for classical and quantum computers.

References

1. Rapid solution of problems by quantum computation. Proc. R. Soc. Lond. A **439**(1907), 553–558 (1992). https://doi.org/10.1098/rspa.1992.0167
2. E. Bernstein, U. Vazirani, Quantum complexity theory. SIAM J. Comput.**26**(5), 1411–1473 (1997). https://doi.org/10.1137/S0097539796300921
3. D.R. Simon, On the power of quantum computation. SIAM J. Comput. **26**(5), 1474–1483 (1997). https://doi.org/10.1137/S0097539796298637
4. P. W. Shor, Algorithms for quantum computation: discrete logarithms and factoring, in *Proceedings 35th Annual Symposium on Foundations of Computer Science* (IEEE Computer Society Press, Santa Fe, NM, USA, 1994), pp. 124–134. https://doi.org/10.1109/SFCS.1994.365700

Part IV

Programming Quantum Computers

Having developed a foundational understanding of quantum computation and its theoretical building blocks, we now turn to the practical task of programming quantum computers. This part focuses on bridging theory and implementation, equipping readers with the tools to construct and run quantum circuits using a modern software framework.

Chapter 16, "The Quantum Computing Stack", introduces the conceptual structure shown in Fig. 63. It discusses hardware-level control, device-specific gate sets, dynamic circuit instructions, and the role of quantum assembly languages. It also provides an overview of error suppression, mitigation, and correction as they appear across the stack.

Chapter 17, "Libraries for Quantum Computing", surveys widely used frameworks and libraries for quantum software development. The chapter includes tools for algorithm prototyping, simulation, classical preprocessing, and cloud-based execution, with an emphasis on Qiskit and supporting libraries.

16 The Quantum Computing Stack

In this chapter, we provide a high-level overview of the programming stack for quantum computing, with examples of various libraries and implementations and brief expositions of strategies for dealing with noise and errors. We first provide an overview of the quantum computer programming stack for gate-based quantum computers in Fig. 16.1. We then explain the components of the stack and their relations and techniques to manage errors and noise in quantum hardware.

We start our discussion from the bottom of the stack and move upwards. The hardware-level controls are specific to the particular implementation of the quantum computers. For superconducting qubits, this is typically the control and shaping of microwave pulses to apply quantum gates and perform measurements [1]. For a photonic quantum computer, this can be an implementation of a phase shifter or a beam splitter [2]. This typically requires detailed knowledge of the physics of hardware implementation. Hardware vendors may provide access to these low-level controls. Quantum circuits allow the abstraction of these details up to at least the hardware gate set provided by vendors.

The device-specific code encompasses the hardware gate set, measurements, and dynamic circuit instructions. Measurements may either be raw samples or expectation values constructed from an ensemble of Pauli strings. The hardware gate set combined with measurements may be considered as the classical equivalent of an instruction set.

Dynamic circuit instructions are quantum instructions conditioned on qubit measurements (classical) within a quantum circuit that are executed at circuit run-time. As an example, using dynamic circuits, if a qubit is measured as 0 during circuit execution, certain gates may be applied; otherwise, measuring 1 leads to other gates being applied. Since quantum devices have short coherence times, dynamic circuit instructions are typically executed on hardware close to the qubits to minimize execution time.

O. M. Raisuddin and S. De, *Quantum Computing for Engineers*,
https://doi.org/10.1007/978-3-032-03325-3_16

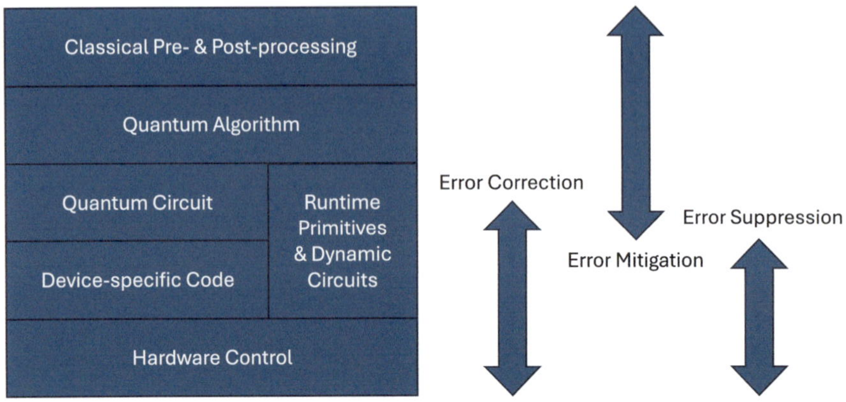

Fig. 16.1 Visualization of the quantum computing stack

Dynamic circuits can allow shortening of quantum circuits in various cases, with two examples being [3, 4]. Since these are more advanced features, they are outside the scope of this text. Error-correction codes rely on dynamic circuits to detect and correct errors on-the-fly.

A quantum assembly language (QASM) is a bridge between device-specific code and the hardware-level implementation of the circuit. OpenQASM [5], cQASM [6], and Jaqal [7] are well-known examples of QASMs. OpenQASM is an open-source language developed by researchers at IBM and the most widely used one in libraries for quantum computing.

A code describing a quantum circuit may be defined using an arbitrary set of gates, which can then be transpiled to code for a target device. In addition to converting to the gate set of the device, the topology of the device must also be considered since logical qubits in the quantum circuit need to be mapped to physical qubits (or a collection of physical qubits corresponding to a logical qubit) in the device. This process is typically executed by a high-level language or library and is referred to as a layout and routing for the tasks of mapping logical qubits to physical qubits, followed by routing operations such as SWAP gates to respect hardware connectivity constraints.

Current quantum devices are in the Noisy Intermediate-Scale Quantum (NISQ) or "pre-fault-tolerant" era. The error rates and coherence times do not meet the threshold required for quantum error correction, and the total number of physical qubits is modest compared to the requirements of error-correcting codes. Quantum error correction is necessary for building fault-tolerant quantum computers which can execute arbitrary quantum algorithms. Details of a few contemporary quantum computing systems with published specifications are listed in Table 16.1. However, we emphasize that such specifications are not fully indicative of the capabilities of the device. As an example, trapped ion qubits have much longer coherence times compared to superconducting chips but the execution times for gates are typically

Table 16.1 Published specifications of gate-based quantum computing hardware developed by major companies

System	Qubit count	Coherence time (T1)	Coherence time (T2)
IBM Heron [8]	133–156	~200 μs	~ 100 μs
Google Willow [9]	105	~68–100 μs	–
IonQ Forte [10]	30–36	~10–100 s	~1 s
Quantinuum H2-1 [11]	56	>60 s	~4 s

longer too. As such, characterizing the performance of quantum computing systems is a nuanced and rapidly evolving subject. For present-day practitioners the critical metric is the maximum 2-qubit gate depth that can be executed on the device without significant error and noise accumulation.

Finally, we note that quantum error correction, mitigation, and suppression can be part of various levels of the quantum computing stack. For classical computers, error correction and fault-tolerance typically occur at the bit level, e.g., voting circuits for logic, redundancy for memory, and parity for communications. In contrast, present-day strategies for dealing with errors for quantum computing range from preprocessing steps of the algorithm to hardware-level controls of individual qubits [12]. Error mitigation and suppression along with other techniques like reordering or optimizing circuits can occur at the quantum circuit or device-specific code levels [13]. Control hardware typically needs to be calibrated periodically to counteract noise and fabrication defects. While all these modalities are vast subjects of research, we provide here a succinct typology to highlight the differences between the three major categories and provide examples of common techniques.

Error Suppression

Quantum error suppression generally refers to techniques for suppressing or reducing the accumulation of errors during the execution of a quantum circuit. While these techniques do not perform error correction, they are an indispensable tool to maximize the performance of NISQ devices. Some commonly used techniques with minimal overhead are dynamical decoupling, Pauli twirling, and twirled readout.

Dynamical decoupling inserts periodic pulses (and their inverse) into idle portions of a quantum circuit to decouple the quantum state from the environment. While this has no net effect in an ideal setting, in the presence of noise these sequences cancel out coupling with the environment in an average sense. This reduces errors and increases the coherence time of qubits. One may experiment with a variety of dynamical decoupling sequences to identify a sequence that works well for the circuit of interest.

Pauli twirling inserts pseudo-random single-qubit Pauli gates into a circuit to convert coherent errors (systematic, unitary errors) into stochastic Pauli errors, which are easier to analyze and mitigate. Pauli twirling can also be applied to two-qubit gates.

Twirled readout modifies the measurement process by flipping a qubit before measurement and then applying a classical NOT operation to the measured result. This effectively averages out certain types of readout bias. Twirling operations are typically inserted randomly in a circuit.

Error Mitigation

Quantum error mitigation techniques are used to improve expectation value estimates by applying a combination of quantum circuit transformations and classical post-processing. The most commonly used technique for error mitigation is zero-noise extrapolation (ZNE). The noise in a quantum circuit is "amplified," and curve fit is used to extrapolate an expectation value to an amplification factor of zero, i.e., zero noise. Various methods can be used to amplify the noise in a circuit. Circuit folding amplifies noise by applying gates (or an entire circuit without measurements) and their inverse. While circuit folding is a conceptually simple method to amplify noise, in practice the circuit depth of a folded circuit typically exceeds the viable circuit depth executable by NISQ hardware. Probabilistic error amplification (PEA) is another common technique for amplifying the error and is better suited for current NISQ hardware. PEA learns the noise in a circuit and amplifies it by explicitly applying the learned noise model at different amplification levels in the circuit. Once expectation values are obtained at different noise amplification levels, a curve fit can be used to extrapolate to an amplification factor of zero. The choice of curve (e.g., linear or quadratic) to be fit to the pairs of expectation value and amplification ratio data points is typically chosen heuristically. For NISQ algorithms, techniques like ZNE are used at the quantum algorithm and pre- and post-processing levels to reduce the effects of noise on measured observables [14].

Error Correction

The prevalent method for constructing logical qubits for fault-tolerant quantum computing is centered around error-correction codes running on a multitude of physical qubits, which run at the device-specific code level. Error-correction codes utilize "syndrome" measurements. While direct measurement of a qubit will lead to the quantum state collapsing to the measured state, a syndrome measurement can be used to test whether two or more qubits have the same quantum state and test for errors without destroying the quantum state. By applying a sequence of syndrome measurements, one may then apply the necessary corrections to restore the quantum state. The theoretical models for error correction have been confirmed experimentally in several experiments [15–17].

While error suppression and error mitigation are useful tools for NISQ devices, error correction is necessary for fault-tolerant quantum computing. Furthermore, the overheads for error mitigation scale exponentially in general for various techniques, which is not a scalable approach towards any possible quantum advantage [18].

Error-corrected, a.k.a. fault-tolerant, quantum computers have not been developed yet. However, as hardware development is progressing, physical qubits may still provide useful results. To improve the performance of physical qubits, error suppression and error mitigation are employed. Error suppression and error mitigation may broadly be differentiated as techniques to reduce errors before measurement, and techniques to reduce the impact of errors after measurement.

References

1. Y. Liu, S.J. Srinivasan, D. Hover, S. Zhu, R. McDermott, A.A. Houck, High fidelity readout of a transmon qubit using a superconducting low-inductance undulatory galvanometer microwave amplifier. New J. Phys. **16**(11), 113008 (2014). https://doi.org/10.1088/1367-2630/16/11/113008
2. P. Kok, W.J. Munro, K. Nemoto, T.C. Ralph, J.P. Dowling, G.J. Milburn, Linear optical quantum computing with photonic qubits. Rev. Mod. Phys. **79**(1), 135–174 (2007). https://doi.org/10.1103/RevModPhys.79.135
3. E. Bäumer et al. Efficient long-range entanglement using dynamic circuits. PRX Quantum **5**(3), 030339 (2024). https://doi.org/10.1103/PRXQuantum.5.030339
4. I. Moflic, A. Paler, On the constant depth implementation of Pauli exponentials (2024). Preprint at arXiv arXiv:2408.08265
5. A. Cross et al., OpenQASM 3: A broader and deeper quantum assembly language. ACM Trans. Quantum Comput. **3**(3), 1–50 (2022). https://doi.org/10.1145/3505636
6. N. Khammassi, G.G. Guerreschi, I. Ashraf, J.W. Hogaboam, C.G. Almudever, K. Bertels, cQASM v1.0: towards a common quantum assembly language (2018). Preprint at arXiv https://doi.org/10.48550/ARXIV.1805.09607
7. B.C.A. Morrison et al., Just another quantum assembly language (Jaqal), in *2020 IEEE International Conference on Quantum Computing and Engineering (QCE)* (IEEE, Denver, CO, USA, 2020), pp. 402–408. https://doi.org/10.1109/QCE49297.2020.00056
8. J. Robledo-Moreno et al., Chemistry beyond exact solutions on a quantum-centric supercomputer (2024). Preprint at arXiv arXiv:2405.05068
9. Google Quantum AI and Collaborators et al., Quantum error correction below the surface code threshold. Nature **638**(8052), 920–926 (2025). https://doi.org/10.1038/s41586-024-08449-y
10. J.S. Chen et al., Benchmarking a trapped-ion quantum computer with 30 qubits. Quantum **8**, 1516 (2024). https://doi.org/10.22331/q-2024-11-07-1516
11. M. DeCross et al., The computational power of random quantum circuits in arbitrary geometries (2024). Preprint at arXiv arXiv:2406.02501
12. S. Günther, N.A. Petersson, J.L. DuBois, Quantum optimal control for pure-state preparation using one initial state. AVS Quantum Sci. **3**(4), 043801 (2021). https://doi.org/10.1116/5.0060262
13. Y. Nam, N.J. Ross, Y. Su, A.M. Childs, D. Maslov, Automated optimization of large quantum circuits with continuous parameters. npj Quantum Inf. **4**(1), 23 (2018). https://doi.org/10.1038/s41534-018-0072-4

14. T. Giurgica-Tiron, Y. Hindy, R. LaRose, A. Mari, W.J. Zeng, Digital zero noise extrapolation for quantum error mitigation, in *2020 IEEE International Conference on Quantum Computing and Engineering (QCE)* (IEEE, Denver, CO, USA, 2020), pp. 306–316. https://doi.org/10.1109/QCE49297.2020.00045
15. S. Bravyi, A.W. Cross, J.M. Gambetta, D. Maslov, P. Rall, T.J. Yoder, High-threshold and low-overhead fault-tolerant quantum memory. Nature **627**(8005), 778–782 (2024). https://doi.org/10.1038/s41586-024-07107-7
16. Google Quantum AI et al., Suppressing quantum errors by scaling a surface code logical qubit. Nature **614**(7949), 676–681 (2023). https://doi.org/10.1038/s41586-022-05434-1
17. Google Quantum AI et al., Exponential suppression of bit or phase errors with cyclic error correction. Nature **595**(7867), 383–387 (2021). https://doi.org/10.1038/s41586-021-03588-y
18. Y. Quek, D. Stilck França, S. Khatri, J.J. Meyer, J. Eisert, Exponentially tighter bounds on limitations of quantum error mitigation. Nat. Phys. **20**(10), 1648–1658 (2024). https://doi.org/10.1038/s41567-024-02536-7

Libraries for Quantum Computing 17

There are various high-level libraries developed for quantum computing, of which the vast majority are available in Python 3 and are open source. Currently, the most advanced and stable library suite is the open-source library Qiskit [1], developed by IBM. Google's Cirq is a Python framework designed for NISQ devices, particularly those based on its Sycamore architecture [2]. While still evolving, it supports a range of algorithms and simulation tools.

Microsoft has taken a different approach with Q# [3], a domain-specific language for quantum computing, tightly integrated with the .NET ecosystem and optimized for hybrid quantum–classical workflows. Other notable libraries are Pennylane [4] and Strawberry Fields [5], developed by Xanadu for their photonic quantum computers, and PyQuil [6] for Rigetti's superconducting qubit architecture.

MATLAB has recently introduced a quantum computing toolbox for constructing and simulating simple quantum circuits and simulating or submitting them to systems available through the cloud. However, it is in a nascent stage and does not offer the high-level functionality available in the Qiskit, Cirq, or Pennylane libraries. The Berkeley Quantum Synthesis Toolkit (BQSkit), developed at Lawrence Berkeley National Laboratory, provides powerful tools for optimizing and synthesizing approximate quantum circuits. TKET is another open-source quantum software developer tool for building and compiling circuits to run on simulators or hardware.

Several of the high-level libraries have built-in implementations of algorithmic primitives, e.g., the quantum Fourier transform or Trotterization. There are also additional libraries and wrappers for implementing algorithms in machine learning and quantum chemistry. TensorFlow Quantum serves as a wrapper that integrates Cirq with TensorFlow, enabling quantum machine learning and variational quantum algorithms. The Qiskit ecosystem includes specialized tools for

O. M. Raisuddin and S. De, *Quantum Computing for Engineers*,
https://doi.org/10.1007/978-3-032-03325-3_17

quantum hardware design through the Qiskit Metal library [1] and software packages for chemistry and quantum physics simulations (Qiskit Nature), numerical methods (Qiskit Algorithms), and device characterization (Qiskit Experiments).

There are several important libraries for the classical preprocessing steps for quantum computation. OpenFermion [7], PySCF [8], Qiskit, and Qiskit Nature can be used for quantum chemistry simulation preprocessing, e.g., for selecting basis sets for molecular structure calculations or implementations of Jordan–Wigner [9] or Bravyi–Kitaev [10] transformations. Quantum signal processing algorithms require a sequence of phase angles to be computed classically. The QSPPACK [11] and PyQSP [12] libraries are the only libraries for this task as of now, with QSPPACK providing superior performance and implementations of state-of-the-art algorithms for phase factor calculations. Since classical preprocessing libraries can vary widely by application area, it is beyond the scope of this book to provide a thorough review of these libraries.

Quantum computers are typically accessed through the cloud. Several commercial, academic, and government research groups have made their prototypes available either directly or through a third party. Notable commercial vendors are IBM, Google, IonQ, Honeywell, Xanadu, and Rigetti, with Amazon bra-ket and Microsoft Azure providing third-party cloud access. Quantum hardware testbeds have been available at Lawrence Berkeley National Laboratory (AQT). Sandia's QSCOUT program has concluded, but insights from it continue to inform future quantum hardware control platforms.

Given Qiskit's comprehensive support across the quantum computing stack, its maturity, and its active development community, we adopt Qiskit as the primary library for programming quantum computers throughout this book.

References

1. Qiskit contributors, *Qiskit: An Open-source Framework for Quantum Computing* (2023). https://doi.org/10.5281/ZENODO.2573505
2. Cirq Developers, *Cirq* (Zenodo, 2023). https://doi.org/10.5281/ZENODO.4062499
3. J. Hooyberghs, *Introducing Microsoft Quantum Computing for Developers: Using the Quantum Development Kit and Q#* (Apress, Berkeley, CA, 2022). https://doi.org/10.1007/978-1-4842-7246-6
4. V. Bergholm et al., PennyLane: automatic differentiation of hybrid quantum-classical computations (2018). https://doi.org/10.48550/ARXIV.1811.04968
5. N. Killoran, J. Izaac, N. Quesada, V. Bergholm, M. Amy, C. Weedbrook, Strawberry fields: a software platform for photonic quantum computing. Quantum **3**, 129 (2019). https://doi.org/10.22331/q-2019-03-11-129
6. P.J. Karalekas et al., *PyQuil: Quantum Programming in Python* (Zenodo, 2020). https://doi.org/10.5281/ZENODO.3553165
7. J.R. McClean et al., OpenFermion: the electronic structure package for quantum computers. Quant. Sci. Technol. **5**(3), 034014 (2020). https://doi.org/10.1088/2058-9565/ab8ebc
8. Q. Sun et al., Recent developments in the PySCF program package. J. Chem. Phys. **153**(2), 024109 (2020). https://doi.org/10.1063/5.0006074
9. P. Jordan, E. Wigner, Über das Paulische äquivalenzverbot. Z. Physik **47**(9–10), 631–651 (1928). https://doi.org/10.1007/BF01331938

10. S.B. Bravyi, A.Y., Kitaev, Fermionic quantum computation. Ann. Phys. **298**(1), 210–226 (2002). https://doi.org/10.1006/aphy.2002.6254
11. Y. Dong, X. Meng, K.B. Whaley, L. Lin, Efficient phase-factor evaluation in quantum signal processing. Phys. Rev. A **103**(4), 042419 (2021). https://doi.org/10.1103/PhysRevA.103.042419
12. J.M. Martyn, Z.M. Rossi, A.K. Tan, I.L. Chuang, Grand unification of quantum algorithms. PRX Quant. **2**(4), 040203 (2021). https://doi.org/10.1103/PRXQuantum.2.040203

Part V
Algorithmic Primitives, Subroutines, and Frameworks

This part introduces a core set of algorithmic building blocks that underpin many of the most powerful quantum algorithms. These primitives enable quantum speedups for diverse classes of problems, including simulation, optimization, and machine learning. The chapters focus on both conceptual understanding and algorithmic structure, with an emphasis on modularity, reusability, and practical implementation.

Chapter 18, "Phase Kickback", illustrates how quantum circuits can encode classical information in the phase of a quantum state, a key mechanism underlying phase estimation and related techniques.

Chapter 19, "Quantum Fourier Transform", presents the quantum analog of the discrete Fourier transform and explains its role in many quantum algorithms, most notably Shor's algorithm and quantum phase estimation.

Chapter 20, "Quantum Phase Estimation", builds on the previous two chapters to show how quantum algorithms can estimate eigenvalues of unitary operators.

Chapter 21, "Trotterization", introduces product formulas that approximate the time evolution of a quantum system—a fundamental technique in quantum simulation.

Chapter 22, "Linear Combination of Unitaries (LCU)", generalizes quantum operations using weighted combinations of unitary operators, enabling implementation of a broader class of linear transformations.

Chapter 23, "Qubitization and Quantum Signal Processing", introduces advanced techniques for operator transformations and polynomial approximations, which underpin modern simulation and linear algebra algorithms.

Chapter 24, "Amplitude Amplification and Estimation", introduces the amplitude amplification subroutine and its application toward Quantum Amplitude Estimation, which are techniques offering quadratic speedups for probabilistic quantum algorithms.

Chapter 25, "Quantum Monte Carlo", presents quantum methods for statistical sampling and expectation estimation, leveraging amplitude estimation for reduced sample complexity.

Chapter 26, "Matrix-Vector Multiplications and Affine Linear Operations", discusses techniques for implementing linear algebraic operations in quantum circuits, including methods for sequences of matrix-vector operations and affine linear matrix-vector operations.

These chapters form the algorithmic core that supports the quantum algorithms introduced in subsequent parts. Each is motivated by practical applications and illustrated with representative code where appropriate.

Phase Kickback

18

Phase kickback is a uniquely quantum phenomenon where the control qubit in a controlled operation accumulates a phase, whereas the target qubit remains unchanged. This is remarkably different from classical computing, where the control bits for logic gates always remain unchanged.

When a controlled gate is applied and the target is in an eigenstate of the unitary being controlled, the corresponding eigenvalue can be "kicked back" to the phase of the control qubit [1]. This effect plays a central role in several quantum subroutines, including quantum phase estimation and the broader quantum signal processing framework.

Consider a unitary operator U with eigenstates $|\lambda_i\rangle$ and corresponding eigenvalues $e^{i\lambda_i}$, such that

$$U = \sum\nolimits_i e^{i\lambda_i} |\lambda_i\rangle\langle\lambda_i|$$

The controlled version of U, denoted by cU, acts as

$$cU = |0\rangle\langle 0| \otimes I + |1\rangle\langle 1| \otimes U$$

Now consider an input state

$$|\psi_i\rangle = \frac{1}{\sqrt{2}}(|0\rangle + |1\rangle) \otimes |\lambda_i\rangle$$

where the first qubit is the control and the second is in the eigenstate of U. Applying cU yields

$$cU|\psi_i\rangle = \frac{1}{\sqrt{2}}\left(|0\rangle + e^{i\lambda_i}|1\rangle\right)$$

O. M. Raisuddin and S. De, *Quantum Computing for Engineers*,
https://doi.org/10.1007/978-3-032-03325-3_18

Thus, the target qubit remains unchanged, while the control qubit picks up a relative phase of $e^{i\lambda_i}$, only for the $|1\rangle$ basis state of the control qubit. This is the essence of phase kickback.

As an example, let's investigate a controlled R_Z gate. The R_Z gate is defined in Qiskit as

$$R_Z(\lambda) = e^{-i\frac{\lambda}{2}Z} = \begin{pmatrix} e^{-i\frac{\lambda}{2}} & 0 \\ 0 & e^{i\frac{\lambda}{2}} \end{pmatrix}$$

This is a diagonal unitary, so its eigenvectors are the computational basis states $|0\rangle$ and $|1\rangle$. Note that other libraries have different, albeit similar, definitions for R_Z gates.

A controlled version, cR_Z has the form:

$$cR_Z(\lambda) = |0\rangle\langle 0| \otimes I + |1\rangle\langle 1| \otimes R_Z(\lambda)$$

which can be expanded in Dirac notation as

$$cR_Z(\lambda) = (|0\rangle\langle 0| \otimes I + |1\rangle\langle 1|) \otimes \left(e^{-i\frac{\lambda}{2}}|0\rangle\langle 0| + e^{i\frac{\lambda}{2}}|1\rangle\langle 1|\right)$$

We examine two cases to illustrate phase kickback:

Case 1: The input is $\frac{1}{\sqrt{2}}(|0\rangle + |1\rangle) \otimes |0\rangle$.

Applying $cR_Z(\lambda)$, we get $\frac{1}{\sqrt{2}}\left(|0\rangle + e^{-i\frac{\lambda}{2}}|1\rangle\right)|0\rangle$.

Case 2: The input is $\frac{1}{\sqrt{2}}(|0\rangle + |1\rangle)|1\rangle$.

The output is $\frac{1}{\sqrt{2}}\left(|0\rangle + e^{i\frac{\lambda}{2}}|1\rangle\right)|1\rangle$.

The following code simulates these two cases and plots the real and imaginary parts of the probability amplitudes of each basis state for $\lambda \in [0, 2\pi]$ in Fig. 18.1:

```
#!/usr/bin/python3


import numpy as np
from matplotlib import pyplot as plt
from qiskit import QuantumCircuit
from qiskit.circuit import Parameter
from qiskit_aer import StatevectorSimulator


# Create parameter lambda
Lambda = Parameter("lambda")


# Create Circuits for the two cases
circ1 = QuantumCircuit(2)
circ1.h(0)
```

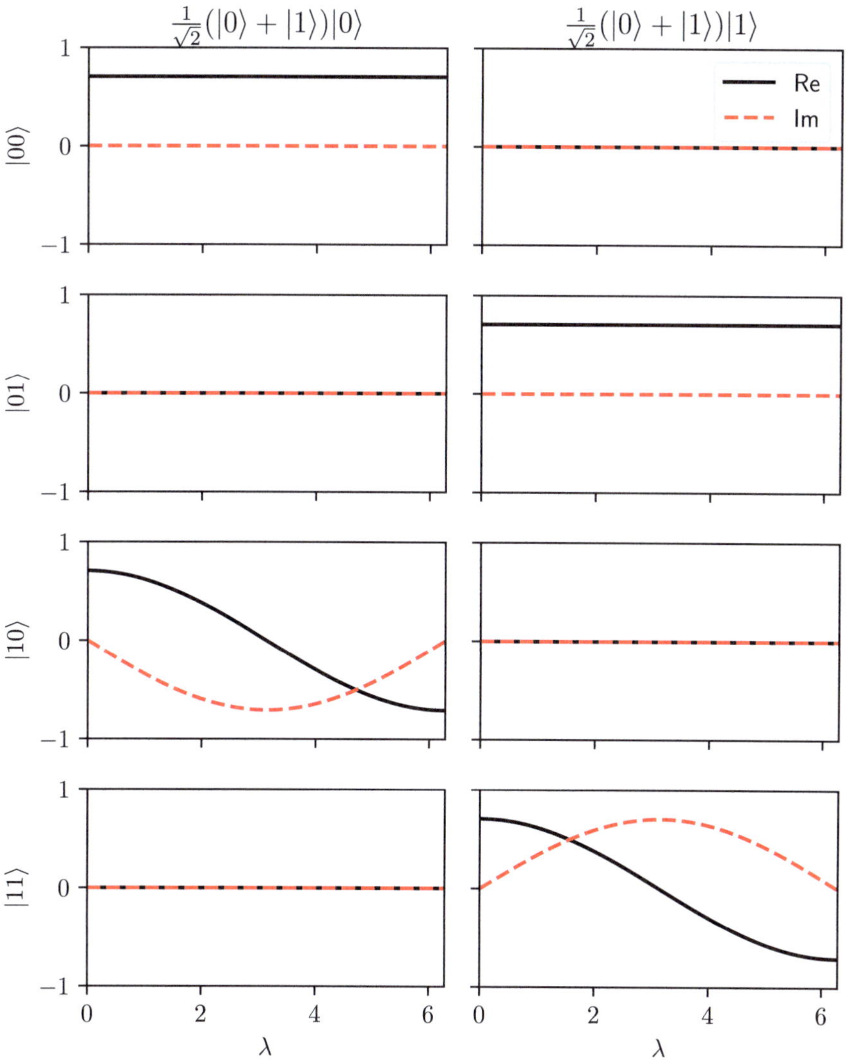

Fig. 18.1 Phase kickback for Left: $cR_Z(\lambda)$ applied to the quantum state $\frac{1}{\sqrt{2}}(|0\rangle + |1\rangle)|0\rangle$; Right: $cR_Z(\lambda)$ applied to the quantum state $\frac{1}{\sqrt{2}}(|0\rangle + |1\rangle)|1\rangle$. The y-axis indicates the probability amplitudes for each basis state (real and imaginary parts shown separately)

```
circ1.crz(Lambda,0,1)

circ2 = QuantumCircuit(2)
circ2.x(1) # Initialize second qubit in state |1>
circ2.h(0)
circ2.crz(Lambda,0,1)
```

```
# Obtain statevectors of the end results for various parameter val-
ues
simulator = StatevectorSimulator()
lambdas = np.linspace(0,np.pi*2,100)
case_1_statevectors   =   np.array([simulator.run(circ1.reverse_
bits().assign_parameters({"lambda":_lambda})).result().get_
statevector() for _lambda in lambdas])
case_2_statevectors   =   np.array([simulator.run(circ2.reverse_
bits().assign_parameters({"lambda":_lambda})).result().get_
statevector() for _lambda in lambdas])

# Plot real and imaginary parts of basis states
plt.rcParams['text.usetex'] = True
fig, axs = plt.subplots(4,2,sharex=True, sharey=True)
for _i in range(4):
  axs[_i,0].plot(lambdas, case_1_statevectors[:,_i].real, 'k-')
  axs[_i,0].plot(lambdas, case_1_statevectors[:,_i].imag, 'r--')
  axs[_i,1].plot(lambdas, case_2_statevectors[:,_i].real, 'k-')
  axs[_i,1].plot(lambdas, case_2_statevectors[:,_i].imag, 'r--')
axs[0,0].set_xlim(0,np.pi*2)
axs[0,0].set_ylim(-1,1)
axs[3,0].set_xlabel(r"$ \lambda $")
axs[3,1].set_xlabel(r"$ \lambda $")
axs[0,1].legend(["Re", "Im"])
axs[0,0].set_ylabel(r"$| 00 \rangle$")
axs[1,0].set_ylabel(r"$| 01 \rangle$")
axs[2,0].set_ylabel(r"$| 10 \rangle$")
axs[3,0].set_ylabel(r"$| 11 \rangle$")
axs[0,0].set_title(r"$ \frac{1}{\sqrt{2}} ( | 0 \rangle + | 1 \ran-
gle ) | 0 \rangle $")
axs[0,1].set_title(r"$ \frac{1}{\sqrt{2}} ( | 0 \rangle + | 1 \ran-
gle ) | 1 \rangle $")
plt.show()
```

We see that the basis states $|0*\rangle$ do not change with lambda, while basis states $|1*\rangle$ pick up a phase of $e^{-i\frac{\lambda}{2}}$ and $e^{i\frac{\lambda}{2}}$ in the left and right plots, respectively.

Reference

1. M.A. Nielsen, I.L. Chuang, *Quantum Computation and Quantum Information: 10th Anniversary Edition*, 1st edn. (Cambridge University Press, 2012). https://doi.org/10.1017/CBO9780511976667

Quantum Fourier Transform 19

The Quantum Fourier Transform (QFT) is the quantum analog of the classical Discrete Fourier Transform (DFT) [1], and it plays a central role in several quantum algorithms, including Shor's factoring algorithm, quantum phase estimation, and quantum signal processing. Unlike the classical DFT, which requires $O(n2^n)$ operations and $O(2^n)$ memory to act on a 2^n-dimensional vector, the QFT can be implemented on an n-qubit quantum computer using only $O(n^2)$ gates (or even $O(n \log n)$ gates with approximation). The QFT transforms the probability amplitudes of quantum states into the Fourier basis, enabling interference-based speedups that are classically intractable.

Formally, given a quantum state $|\psi\rangle$ expressed in the computational basis:

$$|\psi\rangle = \sum_{j=0}^{N-1} \psi_j |j\rangle, \text{ where } N = 2^n.$$

the quantum Fourier transform is the unitary operation $\mathcal{F}_N$ defined by its action on the computational basis states:

$$\mathcal{F}_N |j\rangle = \frac{1}{\sqrt{N}} \sum_{k=0}^{N-1} e^{2\pi i j/N} |k\rangle$$

Applying $\mathcal{F}_N$ to $|\psi$ yields the transformed state:

$$|\phi\rangle = \mathcal{F}_N |\psi\rangle = \sum_{k=0}^{N-1} \phi_k |k\rangle$$

O. M. Raisuddin and S. De, *Quantum Computing for Engineers*,
https://doi.org/10.1007/978-3-032-03325-3_19

where the amplitudes ϕ_k are the discrete Fourier transform of the original amplitudes ψ_j :

$$\phi_k = \frac{1}{\sqrt{N}} \sum_{j=0}^{N-1} \psi_j e^{2\pi ijk/N}$$

To derive a more efficient quantum implementation, we consider the product form of the QFT using the binary representation of indices. For $N = 2^n$, where $n \in \mathbb{Z}^+$, the basis states $|j\rangle \in \{|0\rangle, ..., |2^n - 1\rangle\}$ may be relabeled using the binary notation of the integers j as follows:

$$j = j_1 2^{-1} + j_2 2^{n-2} + \cdots + j_n 2^0 :\rightarrow j_1 j_2 \ldots j_n$$

Similarly, the binary fraction is denoted as

$$j_l/2 + j_l + 1/4 + \cdots + j_m/2^{m-l+1} :\rightarrow 0 \cdot j_l j_{l+1} \ldots j_m$$

Using this notation, the QFT of a basis state $|j\rangle = |j_1 j_2 .. j_n\rangle$ may be written as a product state

$$\mathcal{F}_N |j\rangle = \frac{1}{2^{n/2}} \left(|0\rangle + e^{0.j_n}|1\rangle\right)\left(|0\rangle + \left|e^{0.j_n j_{n-1}}\right\rangle|1\rangle\right) \ldots \left(|0\rangle + e^{0.j_1 j_2 .. j_n}|1\rangle\right)$$

This decomposition reveals that the QFT can be constructed using a sequence of Hadamard and controlled-phase gates and requires only $O(n^2)$ operations.

We provide as an example a Qiskit implementation of a quantum Fourier transform on a uniform superposition of basis states. A uniform superposition of qubits corresponds to an amplitude encoding of a vector of 1's, i.e., $|\Phi\rangle = \frac{1}{\sqrt{N}} \sum_{k=0}^{2^n-1} |k\rangle$, shown as the "before" state in Fig. 1.14, which corresponds to the 0th frequency component in the Fourier basis. Performing a quantum Fourier transform on $|\Phi\rangle$ yields a quantum state $|\Psi\rangle = \mathcal{F}_N|\Phi\rangle$, an amplitude encoding of the Fourier transform of the probability amplitudes of $|\Phi\rangle$. Measuring the qubits after the quantum Fourier transform on $|\Phi\rangle$ yields the quantum state $|\Psi\rangle = |00\ldots0\rangle$, the basis state corresponding to the 0th frequency component as expected, shown in Fig. 19.1 as the "after" state:

```
#!/usr/bin/python3

from matplotlib import pyplot as plt
from qiskit import QuantumCircuit
from qiskit.circuit.library import QFT
from qiskit_aer.primitives import SamplerV2
from qiskit.visualization import plot_histogram
```

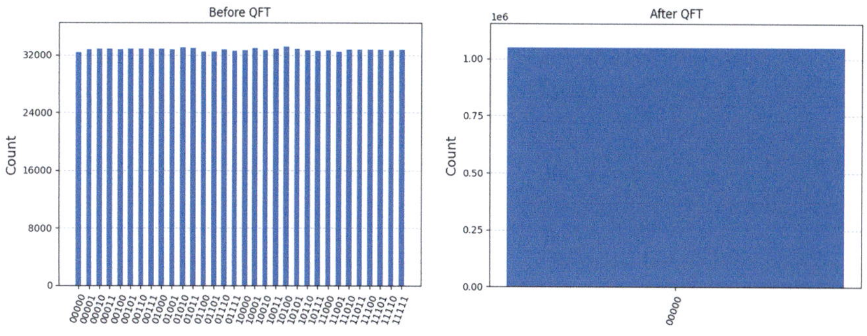

Fig. 19.1 Left: Uniform superposition state to which QFT is applied. Right: Output of QFT; a single basis state

```
beforeFT = QuantumCircuit(5)
# Initialize state in uniform superposition
beforeFT.h([0,1,2,3,4])

# Measure all qubits
beforeFT.measure_all()

afterFT = QuantumCircuit(5)
# Initialize qubits
afterFT.h([0,1,2,3,4])

# Add Fourier transform operation
qft = QFT(num_qubits=5,do_swaps=False).to_gate()
afterFT.append(qft, qargs=[0,1,2,3,4])

# Measure all qubits
afterFT.measure_all()

# Decompose Fourier transform operation into gates for simulator
afterFT = afterFT.decompose(reps=2)

# Simulate the circuit

# Simulate the circuit
sampler = SamplerV2()
job = sampler.run([beforeFT,afterFT],shots=2**20)
result_before = job.result()[0].data.meas.get_counts()
result_after = job.result()[1].data.meas.get_counts()

# Plot a bar chart of all the results
```

```
plot_histogram(result_before,bar_labels=False,title='Before
QFT')
plot_histogram(result_after,bar_labels=False,title='After QFT')

plt.show()
```

Reference

1. C. Coppersmith, An approximate Fourier transform useful in quantum factoring. IBM Research Division, RC 19642 (1994). https://doi.org/10.48550/arXiv.quant-ph/0201067

20 Quantum Phase Estimation

Quantum phase estimation (QPE) is a core subroutine in many quantum algorithms, including Shor's factoring algorithm and expectation value estimation. It provides a way to estimate the eigenvalue associated with a known eigenstate of a unitary operator. At the heart of QPE lies the phenomenon of phase kickback, and the quantum Fourier transform is used to extract the encoded phase information. The algorithm demonstrates one of the key strengths of quantum computing—efficient extraction of global properties (like eigenvalues) from unitary dynamics.

Given an eigenstate $|\lambda_i\rangle$ of a unitary operator U, quantum phase estimation provides an estimate of the corresponding eigenvalue $e^{i2\pi\lambda_i}$. Phase estimation uses the quantum Fourier transform to extract the phase kicked back by a sequence of controlled unitary operations. A series of powers of U, each controlled by a distinct qubit in the clock register, is applied to the eigenstate in the work register. Figure 20.1 shows an example circuit with four control qubits. The control qubits are initialized in uniform superposition before applying the controlled operations. More specifically, the $j\ th$ control qubit applies the unitary U^{2^j} to the work register.

Let us first assume that λ_i can be expressed exactly as a J-bit binary fraction

$$\lambda_i = 0.j_1 j_2 \ldots j_J$$

Denoting $c_j U^{2^j}$ as the controlled-U^{2^j} operations (controlled by the jth qubit in the clock register), the application of the gates on the state:

$$\frac{1}{2^{\frac{J-1}{2}}}(|0\rangle + |1\rangle)^{\otimes J-1} \otimes |\psi\rangle$$

O. M. Raisuddin and S. De, *Quantum Computing for Engineers*,
https://doi.org/10.1007/978-3-032-03325-3_20

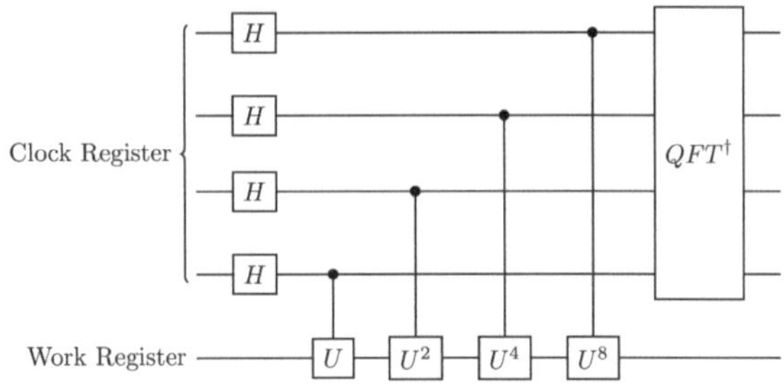

Fig. 20.1 Circuit for quantum phase estimation

yields, via phase kickback:

$$\prod_{j=0}^{J-1}\left(c_j U^{2^j}\right)\left(\frac{1}{2^{\frac{J-1}{2}}}(|0\rangle + |1\rangle)^{\otimes J-1}|\psi\rangle\right)$$
$$= \frac{1}{2^{\frac{J-1}{2}}}\bigotimes_{j=0}^{J-1}\left(|0\rangle + e^{i\lambda_i 2^j}\right)|\psi\rangle$$
$$= \frac{1}{2^{\frac{J-1}{2}}}\sum_{k=0}^{2^{J-1}-1} e^{ik2\pi\lambda_i}|k\rangle|\lambda_m\rangle$$

where $|k\rangle$ ranges over computational basis states of the clock register.

Performing the inverse quantum Fourier transform on the clock register transforms this state into a computational basis state encoding the phase λ_i resulting in $|\lambda_i\rangle$ being sampled with a probability of 1. Note that since $e^{i2\pi\lambda_i}$ is periodic, $|\lambda_i\rangle$ will be measured such that $0 \leq \lambda_i \leq 1$.

Now consider the case where λ_i is not exactly expressible as a J-bit binary fraction, i.e., its closest binary fraction approximation is $\widetilde{\lambda_i}$ *such that.*

$$\lambda_i > \widetilde{\lambda_i} \text{ and } 0 \leq \lambda_i - \widetilde{\lambda_i} \leq \delta$$

Performing phase estimation for this case will not yield a unique $\left|\widetilde{\lambda_i}\right\rangle$. Rather, we will get a probability distribution of sampling various states $|0.j_1 j_2 \ldots j_J\rangle$.

To sample a state from the clock register to obtain λ_i up to m-bits of accuracy, i.e., $|\lambda_i - 0.j_1 j_2 \ldots j_m| = \left|\lambda_i - \overset{m}{\widetilde{\lambda_i}}\right| \leq \frac{1}{2^m}$, we can choose

$$J = m + \left\lceil \log\left(2 + \frac{1}{2\epsilon}\right)\right\rceil$$

To achieve a success probability of $p\left(\left|\lambda_i - \overset{m}{\widetilde{\lambda_i}}\right| \leq \frac{1}{2^m}\right) \geq 1 - \epsilon$.

However, the state $|\lambda_i\rangle$ may not be available or may not be efficiently prepared. In this case, an approximation $\left|\lambda_i'\right\rangle \approx |\lambda_i\rangle$ may be used instead. Expanding this approximation as

$$\left|\lambda_i'\right\rangle = \sum_i c_i|\lambda_i\rangle$$

it can be shown that the probability of measuring $|\widetilde{\lambda_i}\rangle$ is at least

$$p(|\widetilde{\lambda_i}\rangle) = |c_i|^2(1-\epsilon)$$

Here, it must be assumed that $|\lambda_i'\rangle$ has sufficient overlap with $|\widetilde{\lambda_i}\rangle$, i.e., $\langle\lambda_i'|\widetilde{\lambda_l}\rangle$ does not decay exponentially with the number of qubits in the work register. Putting together these results, the complexity of estimating eigenvalues up to a precision ϵ using quantum phase estimation has an overall complexity of $\mathcal{O}(1/\epsilon)$ and has a circuit illustrated in Fig. 20.1.

The standard implementation of the phase estimation algorithm requires J ancilla qubits. A more efficient implementation is the iterative phase estimation algorithm, which requires a single ancilla qubit and mid-circuit measurements to perform phase estimation [1].

We provide the code below to estimate an eigenvalue of the following unitary:

$$U(\phi) = e^{2\pi i\phi Z} = \begin{pmatrix} e^{2\pi i\phi} & 0 \\ 0 & e^{-2\pi i\phi} \end{pmatrix}$$

$U(\phi)$ has an eigenvector $|0\rangle$ with an eigenvalue $e^{2\pi i\phi}$. When quantum phase estimation is applied to this eigenvector, we expect to measure $|\tilde{\phi}\rangle$ with high probability. We implement this in the following code for various values of $-1.5 \leq \phi \leq 1.5$ using 4, 12, and 16 ancilla qubits to demonstrate the behavior of quantum phase estimation and its periodicity, i.e., $\tilde{\phi} \approx \phi$ modulo 1 in Fig. 20.2:

```
#!/usr/bin/python3

import numpy as np
from qiskit.circuit.library import PhaseEstimation
from qiskit.circuit.library import RZGate
from qiskit.primitives import StatevectorSampler
from matplotlib import pyplot as plt

min_ancillae = 8
max_ancillae = 17
step = 4
ancillae = range(min_ancillae,max_ancillae,step)

phi = np.arange(-1.5,1.,0.01).round(3)

phi_measured = np.zeros((len(ancillae),len(phi)))
errors = np.zeros((len(ancillae),len(phi)))

mysampler = StatevectorSampler()
```

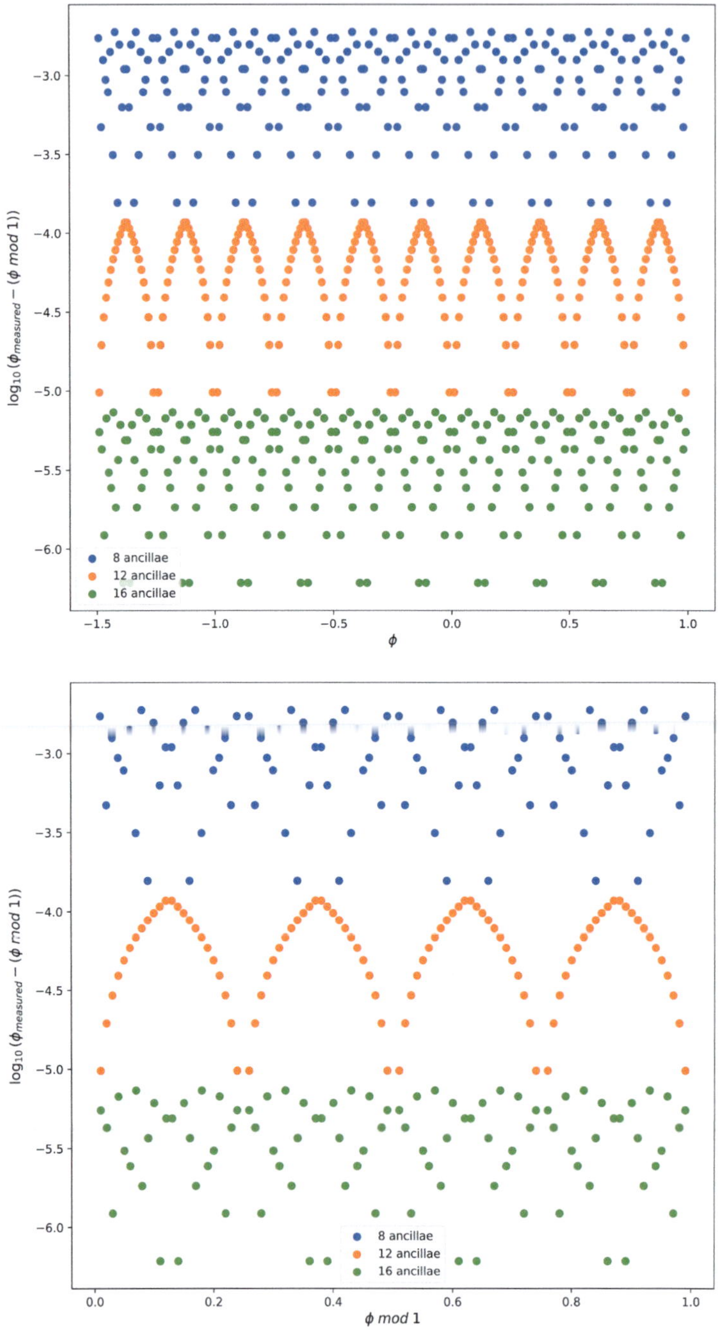

Fig. 20.2 Top: Quantum phase estimation for phi in the range. Bottom: Adjusted plot to account for periodicity of ϕ. Adding 4 qubits to the clock register leads to an effective increase in one digit of precision in the quantum phase estimation of ϕ

```
for i, m in enumerate(ancillae):
  for j in range(len(phi)):
    myunitary = RZGate(-4*np.pi*phi[j])
    myqpe = PhaseEstimation(m,myunitary)
    myqpe.measure_all()
    pub = (myqpe)
    job = mysampler.run([(pub)],shots=1_00_000)
    result = job.result()[0]
    raw = result.data['meas']
    counts = raw.get_counts()
    maxkey = max(counts, key=counts.get)
    # Compute and store the measured value of QPE
    phi_measured[i,j] = 0
    for _i, bit in enumerate(reversed(maxkey)):
    if bit=='1':
    phi_measured[i,j] += 1/(2**(_i+1))
    # Store the errors
    # Phi has a period of 1, adjust for it when computing errors
    errors[i,j] = np.abs((phi[j] % 1) - phi_measured[i,j])

fig, axs = plt.subplots(2)
for i in range(len(ancillae)):
  axs[0].scatter(phi, np.log10(errors[i,:]))
axs[0].set(xlabel='phi', ylabel='log10( phi_measured - (phi mod
1))')
axs[0].legend([str(_i) + ' ancillae' for _i in list(ancillae)])
for i in range(len(ancillae)):
  axs[1].scatter(phi % 1, np.log10(errors[i,:]))
axs[1].set(xlabel='phi mod 1', ylabel='log10( phi_measured - (phi
mod 1))')
axs[1].legend([str(_i) + ' ancillae' for _i in list(ancillae)])
fig.set_size_inches(10, 20)
plt.show()
```

Reference

1. C.J. O'Loan, Iterative phase estimation. J. Phys. A: Math. Theor. **43**(1), 015301 (2010). https://doi.org/10.1088/1751-8113/43/1/015301

21 Trotterization

Trotterization refers to a family of methods that approximate the exponential of a sum of non-commuting operators by products of exponentials of individual terms. This approximation plays a central role in quantum simulation, particularly for simulating time evolution under a Hamiltonian H. It allows us to construct circuits that simulate quantum dynamics without the need for additional ancilla qubits. While not asymptotically optimal, these methods are straightforward to implement and widely used on near-term devices.

Let $H = \sum_{j=0}^{k} H_j$. The Lie–Trotter and Suzuki–Trotter [1] formulas approximate the time evolution operator e^{-iHt} using the Baker–Campbell–Hausdorff formula

$$e^{-iHt} = e^{-i\sum\limits_{j=0}^{k} H_j t} = \left(e^{-i\sum\limits_{j=0}^{k} H_j \frac{t}{r}}\right)^r \approx \left(\prod_{j=0}^{k} e^{-iH_j\frac{t}{r}}\right)^r + O\left(\frac{k^2t^2}{r}\right).$$

Here, r is the number of Trotter steps. The error arises because the matrices H_j generally do not commute, i.e., the commutator $\left[H_j, H_k\right] \neq 0$. Increasing the error in Trotterization arises from the fact that the matrices H_j do not commute in general, i.e., the commutator $\left[H_j, H_k\right] \neq 0$. Increasing r improves the approximation at the cost of longer circuits.

Higher order Trotter formulas can be constructed to reduce the error further. Although the error bounds on Trotter formulas do not scale well, Trotter methods have the advantage of not requiring additional (ancilla) qubits, and the circuit implementations of the individual $e^{iH_j\frac{t}{r}}$ can be straightforward. As an example, the second-order Suzuki–Trotter formula is

$$e^{-iHt} \approx \mathcal{S}_2(t) = \left(\prod_{j=0}^{k} e^{-iH_j\frac{t}{r}} \prod_{j=k}^{0} e^{-iH_j\frac{t}{r}}\right)^r + O\left(\frac{k^3t^3}{r^2}\right).$$

O. M. Raisuddin and S. De, *Quantum Computing for Engineers*,
https://doi.org/10.1007/978-3-032-03325-3_21

More generally, the $(2k)$th order Suzuki–Trotter formula is defined recursively as [2]

$$e^{-iHt} \approx \mathcal{S}_{2k}(t) = \mathcal{S}_{2k-2}(u_k t)^2 \mathcal{S}_{2k-2}((1-4u_k)t)\mathcal{S}_{2k-2}(u_k t)^2$$

where $u_k = \frac{1}{4-4^{1/(2k-1)}}$.

Recent work has shown other higher order formulas with better scaling compared to Trotter–Suzuki formulas [1, 3–5]. The bounds on Trotter errors are known to be loose, but in practice, the errors have been shown to be orders of magnitude lower [6]. Errors in Trotterizations can be reduced by grouping together commuting subsets of $\{H_j\}$ and changing the order of operations in which these groups are applied.

Since H is often decomposed as a sum of Pauli strings, we provide below the procedure and quantum circuits for exponentiating Pauli strings.

For convenience, we express Pauli operators as $\sigma_k \in \{I, X, Y, Z\}$ where $\sigma_0, \sigma_1, \sigma_2, \sigma_3 = I, X, Y, Z$ respectively.

The exponentiation of a single Pauli gate $\sigma_i \in \{X, Y, Z\}$ is straightforward since it is a *local* operator:

$$\exp(-i(I \otimes \ldots \otimes I \otimes \sigma_i \otimes I \otimes \ldots \otimes I)t) = I \otimes \ldots \otimes I \otimes R_i(2t) \otimes I \otimes \ldots \otimes I$$

where $R_i(2t)$ is defined as

$$R_i(2t) = \exp(-it\sigma_i).$$

For Pauli strings describing *nonlocal* operators, i.e., Pauli strings with more than one Pauli gate, the exponentiation is a little more involved. We first describe the kernel for these implementations: the exponentiation of strings of Pauli Z gates.

Consider an arbitrary Pauli string

$$P = \bigotimes_{\substack{j\\k}} \sigma_k$$

Without any loss of generality, let's simply consider the non-trivial portion of the string, i.e., $\sigma_k \neq I$

$$\tilde{P} = \bigotimes_{\substack{j\\k\neq 0}} \sigma_k$$

Similar to the procedure described in Chap. 27, Expectation Value Estimation, for obtaining expectation values, we may diagonalize this $\tilde{P}$ as

$$\tilde{P} = \bigotimes_{\substack{j\\k\neq 0}} V_k^\dagger D_k V_k$$

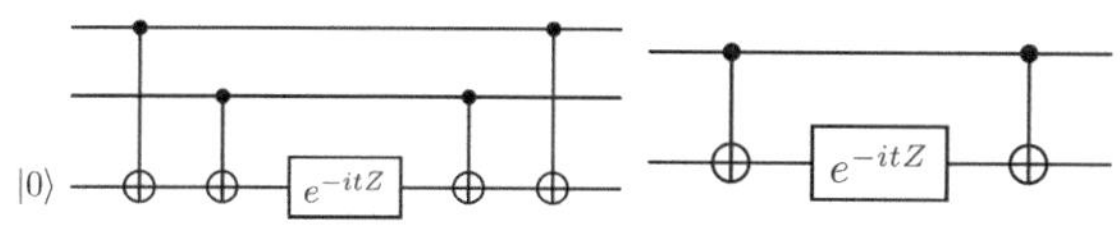

Fig. 21.1 Circuits to exponentiate ZZ using: Left: One ancilla qubit; Right: No ancilla qubits

which can be exponentiated as

$$e^{-i\tilde{P}t} = \bigotimes_{\substack{j \\ k \neq 0}} V_k^\dagger e^{-iD_k t} V_k$$

where $D_k = Z$. Therefore, any arbitrary Pauli string can be exponentiated by applying single-qubit gates $V_k^\dagger$ to transform to the Z basis, exponentiating a Pauli string of Z operators, and transforming back to the computational basis by applying single-qubit gates V_k. Therefore, efficiently exponentiating Pauli Z strings is the kernel of this task.

Two types of techniques have been developed for this operation: ancilla-based and ancilla-free implementations as shown in Fig. 21.1 [7, 8].

The single-qubit Z exponentiation is implemented as $e^{-itZ} = R_Z(2t)$ in Qiskit. These circuits are readily extended to longer Pauli Z strings, and a circuit to exponentiate a general Pauli string $\tilde{P} = \bigotimes_{\substack{j \\ k \neq 0}} \sigma_k$ as $e^{-it\tilde{P}}$ and the cX gates can be arranged as either a "chain/ladder" or as a "fountain" as shown in Fig. 21.2, among other options.

These circuits have been optimized to achieve a depth of $O(\log n)$ [9]. We note that dynamic circuits can further reduce the depth of these circuits from $O(n)$ to $O(1)$ [10].

We provide here as an example a Hamiltonian simulation of single-qubit Pauli terms $H = X + Z$ using the first- and second-order Trotter method for various t, for varying number of Trotter steps, with the error plots shown in Fig. 21.3.

We then provide an example of Hamiltonian simulation of multi-qubit Pauli terms $H = 0.9X \otimes Y \otimes Z + 1.1Y \otimes Z \otimes X$. Since the generation of circuits for multi-qubit Pauli strings is a little more involved, we use built-in functions in Qiskit for this example, with results plotted in Fig. 21.4:

```
#!/usr/bin/python3
import qiskit
from qiskit_aer import UnitarySimulator
```

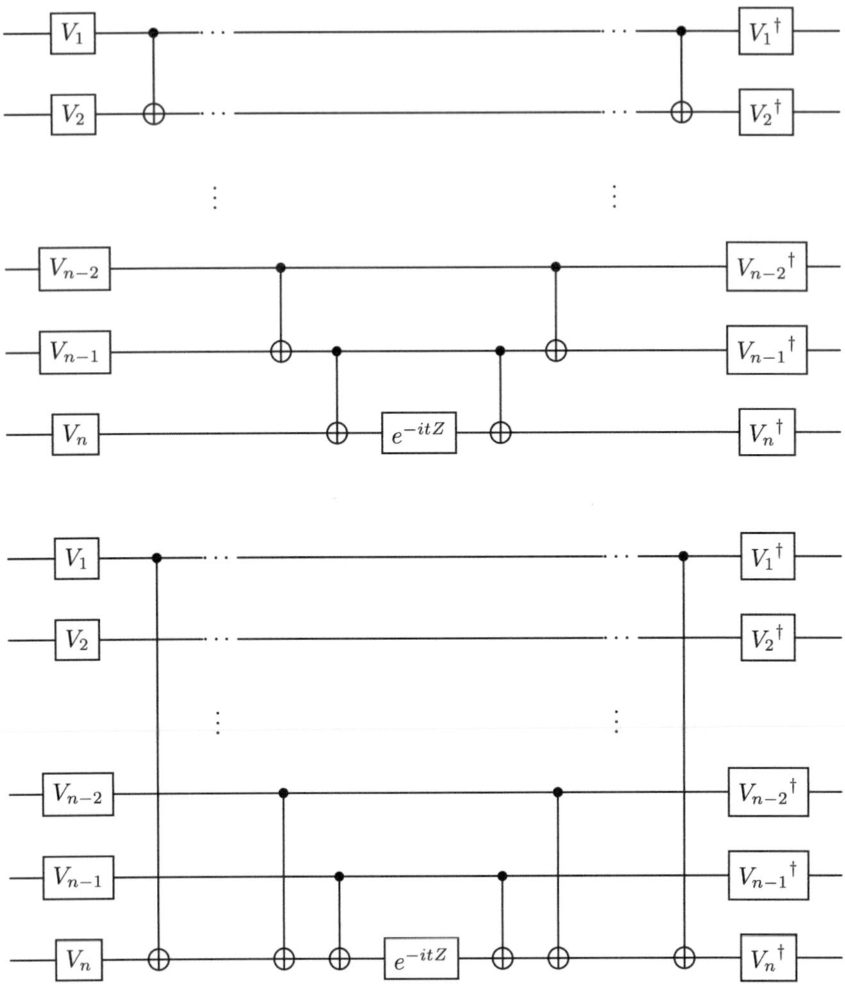

Fig. 21.2 Ancilla-free circuit exponentiating an arbitrary Pauli string $\tilde{P}$ using cX gates arranged as a Top: chain or ladder; Bottom: fountain

```
from scipy.linalg import expm
import numpy as np
import matplotlib.pyplot as plt
min_t = 1
max_t = 5
# Number of Trotter Steps
m = [10,20,40,80,100, 200, 400, 800, 1000, 2000, 4000, 8000, 10000]
# 1st order Trotter
for t in range(min_t,max_t+1):
   # Calculate exact solution classically
```

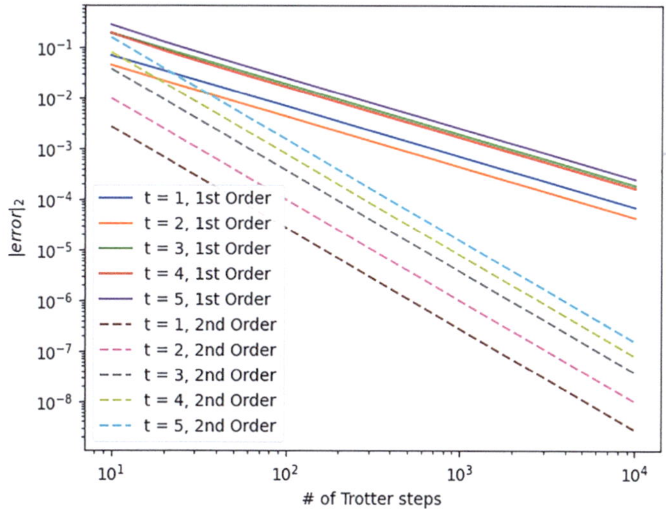

Fig. 21.3 Error scaling of first- and second-order Trotter methods

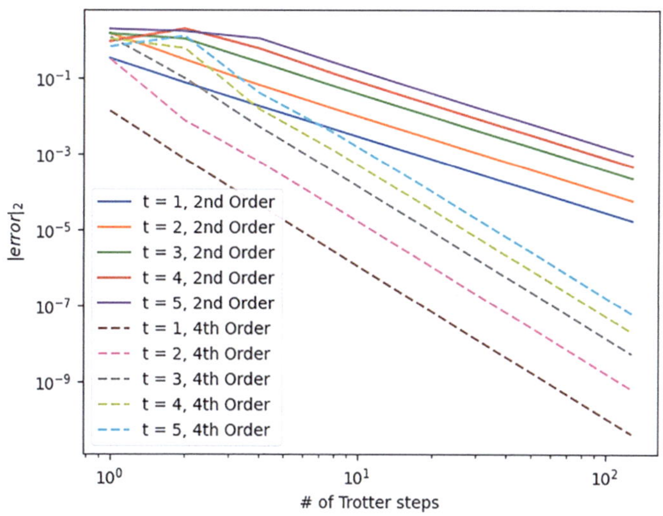

Fig. 21.4 Suzuki–Trotter errors for exponentiating $(0.9X \otimes Y \otimes Z + 1.1Y \otimes Z \otimes X)$ it

```
exact_solution = expm( -t * 1j * ( np.array([[0, 1], [1, 0]])
 + np.array([[1, 0], [0, -1]])))
errors = []
for r in m:
   # Create a register of 1 qubit
   myQRegister = qiskit.QuantumRegister(1, '\psi')
```

```
        # Create a quantum circuit with using myRegister
        myCircuit = qiskit.QuantumCircuit(myQRegister)
        for _r in range(r):
          myCircuit.rx(t*2/r,0)
          myCircuit.rz(t*2/r,0)
        # Simulate the circuit to obtain overall unitary of Trotteri-
zation
        mySimulator = UnitarySimulator()
        result = mySimulator.run(myCircuit). result()
        finalUnitary = result.get_unitary()
        # Compare circuit unitary with exact unitary
              errors.append( np.linalg.norm(finalUnitary - exact_
solution,2))
    plt.loglog(m,errors)
# 2nd order Trotter
for t in range(min_t,max_t+1):
    # Calculate exact solution classically
    exact_solution = expm( -t * 1j * ( np.array([[0, 1], [1, 0]])
     + np.array([[1, 0], [0, -1]])))
    errors = []
    for r in m:
        # Create a register of 1 qubit
        myQRegister = qiskit.QuantumRegister(1, '\psi')
        # Create a quantum circuit with using myRegister
        myCircuit = qiskit.QuantumCircuit(myQRegister)
        for _r in range(r):
          myCircuit.rx(t/r,0)
          myCircuit.rz(t*2/r,0)
          myCircuit.rx(t/r,0)
        # Simulate the circuit to obtain overall unitary of Trotteri-
zation
        mySimulator = UnitarySimulator()
        result = mySimulator.run(myCircuit).result()
        finalUnitary = result.get_unitary()
        # Compare circuit unitary with exact unitary
              errors.append( np.linalg.norm(finalUnitary - exact_
solution,2))
    plt.loglog(m,errors,linestyle='dashed')
plt.xlabel('# of Trotter steps')
plt.ylabel(r'$|error|_2$')
plt.legend(['t = {}, 1st Order'.format(t) for t in range(min_t,max_
t+1)]
+['t = {}, 2nd Order'.format(t) for t in range(min_t,max_t+1)])
plt.show()
```

```
#!/usr/bin/python3
import qiskit
from qiskit_aer import UnitarySimulator
from scipy.linalg import expm
import numpy as np
import matplotlib.pyplot as plt
from qiskit.quantum_info import SparsePauliOp
from qiskit.circuit.library import PauliEvolutionGate
from qiskit.synthesis import SuzukiTrotter
min_t = 1
max_t = 5
# Number of Trotter Steps
m = [1,2,4,8,16,32,64,128]
myOp = SparsePauliOp("XYZ",0.9) + SparsePauliOp("YZX",1.1)
# 2nd order Suzuki-Trotter
for t in range(min_t,max_t+1):
  # Calculate exact solution classically
  exact_solution = expm( -t * 1j * myOp.to_matrix())
  errors = []
  for r in m:
    # Define Suzuki-Trotter method
          mySynthesis = SuzukiTrotter(order=2, reps=r, cx_
structure='chain')
    # Create Pauli Evolution
      myEvolution = PauliEvolutionGate(myOp,time=t,synthesis=
mySynthesis)
    # Append to a quantum circuit
    myCircuit = qiskit.QuantumCircuit(myOp.num_qubits)
    myCircuit.append(myEvolution,range(0,myOp.num_qubits))
    # Simulate the circuit to obtain overall unitary of Trotteri-
zation
    mySimulator = UnitarySimulator()
       result = mySimulator.run(myCircuit.decompose(reps=2)).
result()
    finalUnitary = result.get_unitary()
    # Compare circuit unitary with exact unitary
         errors.append( np.linalg.norm(finalUnitary - exact_
solution,2))
  plt.loglog(m,errors)
# 4th order Suzuki-Trotter
for t in range(min_t,max_t+1):
  # Calculate exact solution classically
```

```
    exact_solution = expm( -t * 1j * myOp.to_matrix())
    errors = []
    for r in m:
        # Define Suzuki-Trotter method
        mySynthesis = SuzukiTrotter(order=4, reps=r, cx_structure='chain')
        # Create Pauli Evolution
        myEvolution = PauliEvolutionGate(myOp,time=t,synthesis=mySynthesis)
        # Append to a quantum circuit
        myCircuit = qiskit.QuantumCircuit(myOp.num_qubits)
        myCircuit.append(myEvolution,range(0,myOp.num_qubits))
        # Simulate the circuit to obtain overall unitary of Trotterization
        mySimulator = UnitarySimulator()
        result = mySimulator.run(myCircuit.decompose(reps=2)).result()
        finalUnitary = result.get_unitary()
        # Compare circuit unitary with exact unitary
        errors.append( np.linalg.norm(finalUnitary - exact_solution,2))
    plt.loglog(m,errors,linestyle='dashed')
plt.xlabel('# of Trotter steps')
plt.ylabel(r'$|error|_2$')
plt.legend(['t = {}, 2nd Order'.format(t) for t in range (min_t,max_t+1)]
+['t = {}, 4th Order'.format(t) for t in range(min_t,max_t+1)])
plt.show()
```

References

1. M. Suzuki, Generalized Trotter's formula and systematic approximants of exponential operators and inner derivations with applications to many-body problems. Commun. Math. Phys. **51**(2), 183–190 (1976). https://doi.org/10.1007/BF01609348
2. M. Suzuki, General theory of higher-order decomposition of exponential operators and symplectic integrators. Phys. Lett. A **165**(5–6), 387–395 (1992). https://doi.org/10.1016/0375-9601(92)90335-J
3. E. Remez, Sur le calcul effectif des polynomes d'approximation de tchebichef. CR Acad. Sci. Paris (1934)
4. C.-H. Cho, D.W. Berry, M.-H. Hsieh, Doubling the order of approximation via the randomized product formula. Phys. Rev. A **109**(6), 062431 (2024). https://doi.org/10.1103/PhysRevA.109.062431
5. M.E.S. Morales, P.C.S. Costa, D.K. Burgarth, Y.R. Sanders, D.W. Berry, Greatly improved higher-order product formulae for quantum simulation (2022). https://doi.org/10.48550/ARXIV.2210.15817

6. A.M. Childs, D. Maslov, Y. Nam, N.J. Ross, Y. Su, Toward the first quantum simulation with quantum speedup. Proc. Natl. Acad. Sci. U.S.A. **115**(38), 9456–9461 (2018). https://doi.org/10.1073/pnas.1801723115
7. M.A. Nielsen, I.L. Chuang, *Quantum Computation and Quantum Information: 10th Anniversary Edition*, 1st edn. (Cambridge University Press, 2012). https://doi.org/10.1017/CBO9780511976667
8. J.D. Whitfield, J. Biamonte, A. Aspuru-Guzik, Simulation of electronic structure Hamiltonians using quantum computers. Mol. Phys. **109**(5), 735–750 (2011). https://doi.org/10.1080/00268976.2011.552441
9. A. Cowtan, S. Dilkes, R. Duncan, W. Simmons, S. Sivarajah, Phase gadget synthesis for shallow circuits. Electron. Proc. Theor. Comput. Sci. **318**, 213–228 (2020). https://doi.org/10.4204/EPTCS.318.13
10. I. Moflic, A. Paler, On the Constant Depth Implementation of Pauli Exponentials (2024). https://doi.org/10.48550/arXiv.2408.08265. arXiv:2408.08265

Linear Combination of Unitaries

22

Many quantum algorithms require the application of an operator that is not directly unitary, but rather a linear combination of known unitaries. This situation arises, for instance, in Hamiltonian simulation, quantum singular value transformation, and quantum linear systems solvers. The linear combination of unitaries (LCU) method enables such operators to be implemented on a quantum computer, assuming circuit-level access to each unitary component. It is especially powerful because it allows one to process non-Hermitian or even general matrices when combined with block-encoding and qubitization techniques [1].

Suppose we are given circuits or oracles for the unitaries U_l and their associated coefficients α_l. We want to implement the operator

$$U = \sum_l \alpha_l U_l$$

on a quantum state. The LCU method accomplishes this by combining two subroutines: *SELECT* and *PREPARE*.

The *PREPARE* operation encodes the coefficients into a quantum superposition:

$$PREPARE|0\rangle^{\otimes m} = |\alpha\rangle$$

where $|\alpha\rangle = \frac{1}{\alpha_1} \sum_l \sqrt{\alpha_l}|l\rangle$ s.t. $\alpha_l \in \mathbb{R}^+$. This is without any loss of generality since any phase of the coefficients can be absorbed into the unitary.

The *SELECT* operation applies the appropriate unitary based on the index in the ancilla register:

$$SELECT = \sum_l |l\rangle\langle l| \otimes U_l$$

O. M. Raisuddin and S. De, *Quantum Computing for Engineers*,
https://doi.org/10.1007/978-3-032-03325-3_22

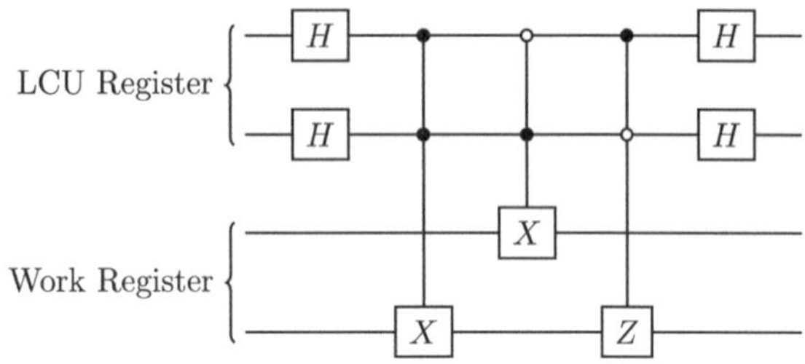

Fig. 22.1 A circuit implementing the linear combination of unitaries of Pauli matrices in the example and code of this chapter

To implement the linear combination of unitaries, the following sequence is used:

$$\left(PREPARE^{\dagger} \otimes I\right)(SELECT)(PREPARE \otimes I).$$

Measuring the ancilla register in the state $|0\rangle^{\otimes m}$ indicates the successful application of the linear combination of unitaries.

Note that the LCU method is an instance of block encoding of an arbitrary matrix H, where H can be non-Hermitian, into a larger unitary. Block-encoded access to a matrix enables the use of powerful quantum algorithmic frameworks, including qubitization, introduced in Chap. 23, Qubitization and Quantum Signal Processing. A particularly efficient and versatile application of qubitization is quantum signal processing (QSP).

Finally, we note that the LCU method can be used to add two matrices H_1 and H_2 via their block-encodings U_{H_1} and U_{H_2}.

We provide as an example a Qiskit implementation of a circuit taking a linear combination of the unitaries:

$$I \otimes I + X \otimes I + I \otimes X + I \otimes Z = \begin{pmatrix} 2 & 1 & 1 & 0 \\ 1 & 0 & 0 & 1 \\ 1 & 0 & 2 & 1 \\ 0 & 1 & 1 & 0 \end{pmatrix}$$

where the coefficients $\alpha_l = \frac{1}{4} \; \forall \; l \in [0, 3]$, for which the corresponding *PREPARE* operation is simply a uniform superposition achieved by Hadamard gates. We provide an implementation below with the corresponding circuit shown in Fig. 22.1:

```
#!/usr/bin/python3


import qiskit
from qiskit_aer import UnitarySimulator
from qiskit.circuit.library import XGate, ZGate
import numpy as np
```

```
# LCU register
lcuRegister = qiskit.QuantumRegister(2, 'LCU')
# Work register
workRegister = qiskit.QuantumRegister(2, '\psi')

myCircuit = qiskit.QuantumCircuit(workRegister,lcuRegister)

# PREP+ operation
myCircuit.h(lcuRegister)

# SELECT operation
myCircuit.append(XGate().control(num_ctrl_qubits=2,ctrl_
state='11'),[*lcuRegister, workRegister[0]])
myCircuit.append(XGate().control(num_ctrl_qubits=2,ctrl_
state='01'),[*lcuRegister, workRegister[1]])
myCircuit.append(ZGate().control(num_ctrl_qubits=2,ctrl_
state='10'),[*lcuRegister, workRegister[0]])

# PREP operation
myCircuit.h(lcuRegister)

# Simulate circuit
mySimulator = UnitarySimulator()
result = mySimulator.run(myCircuit.decompose(reps=2)).result()
# Extract subspace of successfully measuring LCU qubits as 0
# and multiply by submormalization factor 4
print(np.array(result.get_unitary().data[0:4,0:4]).round(10)*4)
```

Reference

1. A.M. Childs, N. Wiebe, Hamiltonian simulation using linear combinations of unitary operations. QIC **12**(11, 12) (2012). https://doi.org/10.26421/QIC12.11-12

Qubitization and Quantum Signal Processing

23

Many advanced quantum algorithms rely on applying polynomial transformations to the eigenvalues or singular values of a matrix. Such transformations are foundational to quantum algorithms for Hamiltonian simulation, linear system solving, amplitude amplification, and quantum machine learning. Two powerful frameworks—Qubitization and Quantum Signal Processing (QSP)—enable these polynomial transformations efficiently and with provable optimality in terms of gate complexity.

Qubitization

Qubitization constructs a unitary operator that acts as an $SU(2)$ rotation in a 2D subspace associated with each eigenvalue of a matrix A [1]. This allows the application of Chebyshev polynomial transformations to eigenvalues using repeated applications of a structured unitary.

Consider the 2×2 rotation matrix:

$$O(\lambda) = \begin{pmatrix} \lambda & -\sqrt{1-\lambda^2} \\ \sqrt{1-\lambda^2} & \lambda \end{pmatrix}.$$

whose powers are given by

$$O^k(\lambda) = \begin{pmatrix} T_k(\lambda) & -\sqrt{1-\lambda^2}U_{k-1}(\lambda) \\ \sqrt{1-\lambda^2}U_{k-1}(\lambda) & T_k(\lambda) \end{pmatrix}$$

O. M. Raisuddin and S. De, *Quantum Computing for Engineers*,
https://doi.org/10.1007/978-3-032-03325-3_23

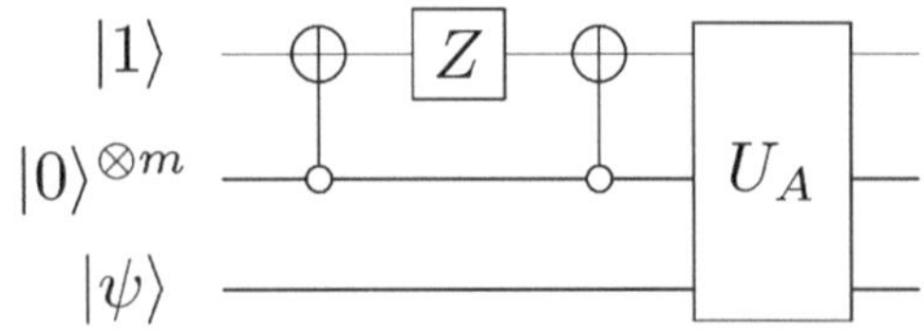

Fig. 23.1 The operation Z_Π followed by U_A

where $T_k(\lambda)$ and $U_k(\lambda)$ are Chebyshev polynomials of the first and second kind, respectively. If the scalar λ is replaced with a matrix A, then the matrix

$$O^k(A) = \begin{pmatrix} T_k(A) & -\sqrt{1-A^2}U_{k-1}(A) \\ \sqrt{1-A^2}U_{k-1}(A) & T_k(A) \end{pmatrix}$$

is a block-encoding of $T_k(A)$. Using a linear combination of these Chebyshev polynomials (using LCU techniques introduced in Chap. 22: Linear Combination of Unitaries), any arbitrary polynomial of the matrix Awhich is the desired equivalent can be created.

To extend this to arbitrary block-encodings of the form:

$$U_A = \begin{pmatrix} A & * \\ * & * \end{pmatrix}; \text{ where } A = \langle 0^{\otimes m}|U_A|0^{\otimes m}\rangle$$

qubitization introduces a reflection:

$$Z_\Pi = 2|0^{\otimes m}\, 0^{\otimes m}| \otimes I - I$$

to apply the operation $O_A = U_A Z_\Pi$, which is the desired equivalent of $O(A)$, as shown in the circuit in Fig. 23.1.

For a Hermitian A, repeated application of $O_A = U_A Z_\Pi$ will yield the desired Chebyshev polynomial of A. For non-Hermitian A, the alternating sequence $O_A = U_A Z_\Pi U_A^\dagger Z_\Pi$ can be used instead.

Since the block-encoding of the resulting Chebyshev polynomial (an orthogonal polynomial basis) is a unitary operation, any arbitrary polynomial can be constructed using a linear combination of these unitaries. However, this would require an overhead of up to $O(\log(d))$ ancillary qubits where d is the degree of the polynomial.

Quantum Signal Processing

Quantum signal processing (QSP) [2] addresses the issue of multiple ancilla qubits required to form arbitrary polynomials as linear combinations of unitaries. This is achieved by forming a polynomial directly by applying rotation operations to the ancilla qubit. The main idea behind quantum signal processing is to replace the Z gate in qubitization with the general Z rotations $R_Z(2\phi_i)$, where $\phi_i \in \Phi \in \mathbb{R}^{d+1}$

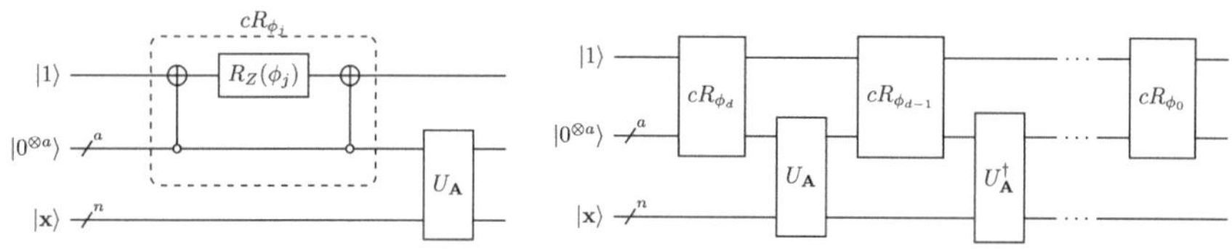

Fig. 23.2 Top: The controlled rotation operation cR_{ϕ_j}. Bottom: A quantum signal processing circuit for a non-Hermitian block-encoding

is a predetermined sequence of $d+1$ rotation angles to form a block-encoding of the desired polynomial $P(A)$:

$$P(A) = \langle 0|^{\otimes m} U_\Phi(A) |0\rangle^{\otimes m}$$

$$U_\Phi(A) = e^{-i\phi_0 Z} \prod_{j=1}^{d} \left[U_A e^{-i\phi_j Z} \right]$$

where A is Hermitian. An overview of a quantum signal processing circuit for a non-Hermitian block-encoding U_A of a matrix A is given in Fig. 23.2, where block-encoding and its Hermitian conjugate are used in an alternating sequence. The quantum signal processing theorem [2] states that a sequence of $d+1$ rotation angles exists for any complex polynomial P of maximum degree d, and the polynomial is even for even d and odd for odd d. Furthermore, it states that the complex conjugate of the polynomial can be formed using the phase angles $-\Phi$.

Theorem (Quantum Signal Processing [2]) *The quantum signal processing sequence U_Φ produces a matrix that may be expressed as a polynomial function of x:*

$$U_\Phi = e^{i\phi_0 Z} \prod_{k=1}^{d} W(x) e^{i\phi_k Z} = \begin{pmatrix} P(a) & iQ(a)\sqrt{1-a^2} \\ iQ^\dagger(a)\sqrt{1-a^2} & P^*(a) \end{pmatrix}$$

For $x \in [-1, 1]$, and a Φ exists for any polynomials P, Q in x s.t.:

(i) $\deg(P) \leq d, \deg(Q) \leq d-1$.
(ii) P has a parity (is even or odd) d mod 2 and Q has a parity $(d-1)$ mod 2.
(iii) $|P|^2 + \left(1-a^2\right)\left|Q^2\right| = 1$. □

Using this result, we note that any arbitrary polynomial can be formed by taking a linear combination of the even and odd parts of the polynomial. Furthermore, the real or complex parts of a polynomial can be extracted by taking a linear combination of the polynomial and its complex conjugate. There are various libraries for calculating the required phase angles for any desired polynomial [2, 3], with

QSPPACK [3] providing state-of-the-art performance at the time of writing this manuscript.

Recent work dubbed Generalized Quantum Signal Processing has generalized the Z rotations on the ancilla qubit to include X and Y rotations [4], which allows arbitrary polynomials of degree d to be constructed without parity decompositions. This method requires access to a block-encoding of e^{iA}, rather than A itself.

Quantum signal processing is currently the most powerful and optimal method for many quantum algorithms. The quantum eigenvalue transform, quantum singular value transform [5], factoring, phase estimation, Hamiltonian simulation, linear system solution, amplitude amplification, eigenstate filtering, and many other quantum computing problems can be reformulated as a quantum signal processing problem with optimal or near-optimal scaling results [2].

In Fig. 23.2, we show the implementation of a controlled rotation using a phase angle ϕ_i and a quantum signal processing circuit implementing the sequence of controlled rotations Φ to implement a polynomial of a matrix A using its block-encoding U_A.

Quantum Eigenvalue Transformation and Quantum Singular Value Transformation

We can now summarize the quantum signal processing procedure applied to Hermitian and non-Hermitian matrices as general frameworks referred to as Quantum Eigenvalue Transformation and Quantum Singular Value Transformation.

For a Hermitian matrix $A = V^\dagger \Lambda V$, applying quantum signal processing directly yields

$$P(A) = V^\dagger P(\Lambda) V$$

which is known as the *quantum eigenvalue transformation.*

For a non-Hermitian matrix $A = U\Sigma V^\dagger$, one can form odd polynomials

$$P_{odd}(A) = UP_{odd}(\Sigma)V^\dagger$$

and even polynomials

$$P_{even}(A) = V^\dagger P_{even}(\Sigma)V^\dagger \text{ or } P_{even}(A) = UP_{even}(\Sigma)U$$

by alternating between U_A and $U_A^\dagger$. This is known as the *quantum singular value transformation.*

Though being unable to form even polynomials of A of the form $UP_{even}(\Sigma)V^\dagger$ may seem restrictive, several important applications can be realized through this framework, e.g., solving linear systems by approximating the odd function x^{-1} with an odd polynomial $P_{x^{-1}}(A) \approx A^{-1}$.

References

1. G.H. Low, I.L. Chuang, Hamiltonian simulation by qubitization. Quantum **3**, 163 (2019). https://doi.org/10.22331/q-2019-07-12-163
2. J.M. Martyn, Z.M. Rossi, A.K. Tan, I.L. Chuang, Grand unification of quantum algorithms. PRX Quant. **2**(4), 040203 (2021). https://doi.org/10.1103/PRXQuantum.2.040203
3. Y. Dong, X. Meng, K.B. Whaley, L. Lin, Efficient phase-factor evaluation in quantum signal processing. Phys. Rev. A **103**(4), 042419 (2021). https://doi.org/10.1103/PhysRevA.103.042419
4. D. Motlagh, N. Wiebe, Generalized quantum signal processing (2023). *arXiv*. https://doi.org/10.48550/ARXIV.2308.01501
5. A. Gilyén, Y. Su, G.H. Low, N. Wiebe, Quantum singular value transformation and beyond: exponential improvements for quantum matrix arithmetics, in *Proceedings of the 51st Annual ACM SIGACT Symposium on Theory of Computing* (ACM, Phoenix AZ USA, 2019), pp. 193–204. https://doi.org/10.1145/3313276.3316366

24 Amplitude Amplification and Estimation

Many quantum algorithms, including Grover's search and quantum Monte Carlo, are inherently probabilistic and produce a desired outcome with some success probability $p < 1$. When this success probability is low, repeating the algorithm naively increases the number of required samples to $O(1/p)$, which can be inefficient. Amplitude amplification is a powerful quantum technique that boosts this success probability quadratically, requiring only $O\left(1/\sqrt{p}\right)$ repetitions [1]. When p is unknown, amplitude estimation provides an efficient way to estimate it using elements of phase estimation, again achieving quadratic speedup over classical sampling methods.

Quantum Amplitude Amplification

Consider a quantum algorithm $\mathcal{A}$ acting on the input state $|0\rangle^{\otimes n}$ to prepare an output state

$$\mathcal{A}|0\rangle^{\otimes n} = |\psi\rangle = \sqrt{p}\left|\psi_{good}\right\rangle + \sqrt{1-p}|\psi_{bad}\rangle$$

where

$$\left\langle\psi_{good}|\psi\right\rangle = \left\langle\psi_{good}|\psi_{good}\right\rangle = p$$

$$\langle\psi_{bad}|\psi\rangle = \langle\psi_{bad}|\psi_{bad}\rangle = 1 - p$$

s.t. $\left|\psi_{good}\right\rangle$ is the desired output with success probability p. The amplitude amplification procedure will boost the success probability to $O(1)$ using $O\left(1/\sqrt{p}\right)$ applications of $\mathcal{A}$.

O. M. Raisuddin and S. De, *Quantum Computing for Engineers*,
https://doi.org/10.1007/978-3-032-03325-3_24

More specifically, this is achieved using $O(\sqrt{p})$ applications of an operator $\mathcal{Q}$, sometimes referred to as a Grover operator, of the form:

$$\mathcal{Q} = -\mathcal{A}\mathcal{S}_0\mathcal{A}^{-1}\mathcal{S}_{good}$$

Here, $\mathcal{S}_{good}$ is an operator that marks the good subspace with a −ve phase:

$$\mathcal{S}_{good}|\psi\rangle = -\sqrt{p}|\psi_{good}\rangle + \sqrt{1-p}|\psi_{bad}\rangle$$

and $\mathcal{S}_0$ is an operator that marks the state $|0\rangle^{\otimes n}$ with a −ve sign:

$$\mathcal{S}_0\left(\sqrt{p}|0\rangle^{\otimes n} + \sqrt{1-p}|\perp\rangle\right) = -p|0\rangle^{\otimes n} + \sqrt{1-p}|\perp\rangle$$

Note that the operator $\mathcal{S}_{good}$ is specific to $\mathcal{A}$ [2] provides a guide to implementing the operator $\mathcal{S}_{good}$ given $\mathcal{A}$. On the other hand, the circuit shown in Fig. 24.1 is an implementation of the operation $\mathcal{S}_0$, which has the following matrix representation:

$$\mathcal{S}_0 = \begin{pmatrix} -1 & & & & & & \\ & 1 & & & & & \\ & & 1 & & & & \\ & & & \ddots & & & \\ & & & & 1 & & \\ & & & & & 1 & \\ & & & & & & 1 \end{pmatrix}$$

Let us analyze the Grover operator and its effect. First, let us define a 2D subspace spanned by the basis states $|\psi_{good}\rangle$ and $|\psi_{bad}\rangle$. This can be visualized using Fig. 24.2.

Using this figure, one may easily visualize that the operator $-\mathcal{S}_{good}$ simply reflects a state across $|\psi_{bad}\rangle$ as shown in Fig. 24.3.

Therefore, we may rewrite it as a Householder reflector

$$-\mathcal{S}_{good} = I - 2|\psi_{bad}\rangle\langle\psi_{bad}|$$

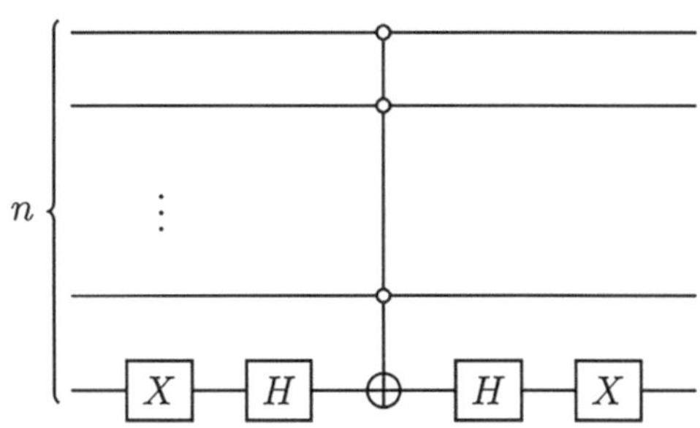

Fig. 24.1 Implementation of $\mathcal{S}_0$, a unitary circuit that marks the state $|0\rangle^{\otimes n}$ with −ve phase

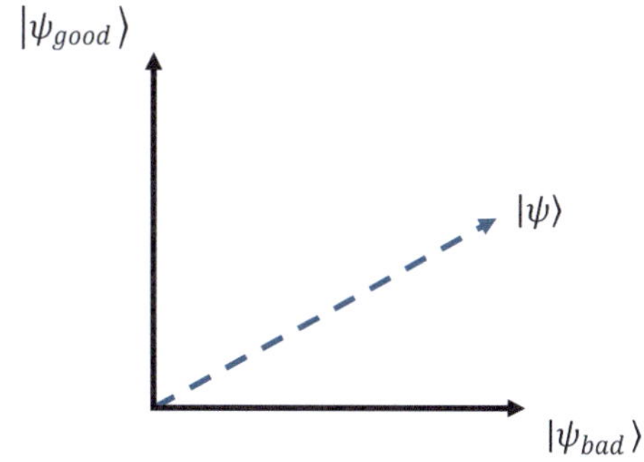

Fig. 24.2 2D subspace spanned by the good and bad states

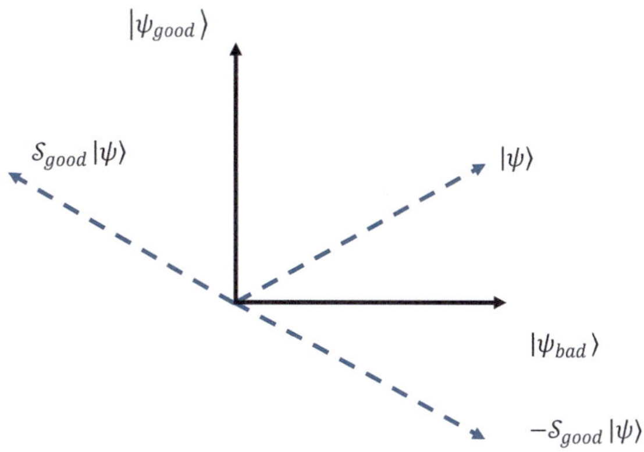

Fig. 24.3 Reflecting across $|\psi_{bad}\rangle$

We now turn to the remaining part of the Grover operator: $\mathcal{A}S_0\mathcal{A}^{-1}$.

Since the operator $\mathcal{S}_0$ simply marks the standard basis state $|0\rangle$ with a $-$ve phase and leaves the rest of the basis states unchanged, it can also be rewritten as the Householder reflector:

$$\mathcal{S}_0 = I - 2|0\rangle\langle 0|$$

Substituting this into $\mathcal{A}S_0\mathcal{A}^{-1}$ we get

$$\begin{aligned}\mathcal{A}S_0\mathcal{A}^{-1} &= \mathcal{A}(I - 2|0\rangle\langle 0|)\mathcal{A}^{-1} \\ &= I - 2|\psi\rangle\langle\psi|\end{aligned}$$

which is clearly a Householder reflector across that state $|\psi\rangle$ in the $\{|\psi_{good}\rangle, |\psi_{bad}\rangle\}$ basis. Using this, we can rewrite the Grover operator as a product of two Householder reflectors

$$\mathcal{Q} = R_{|\psi\rangle}R_{bad}$$

where

$$R_{bad} = I - 2|\psi_{bad}\rangle\langle\psi_{bad}|$$

and

$$R_{|\psi\rangle} = I - 2|\psi\rangle\langle\psi|$$

Geometrically, we may now visualize these two reflections using Fig. 24.4.

Visually we can see that the amplitude of $|\psi\rangle$ has been boosted. We can use this visual argument to rewrite $|\psi\rangle$ and $\mathcal{Q}$:

$$|\psi\rangle = \sin\theta|\psi_{good}\rangle + \cos\theta|\psi_{bad}\rangle$$

$$\mathcal{Q}|\psi\rangle = \sin 3\theta|\psi_{good}\rangle + \cos 3\theta|\psi_{bad}\rangle$$

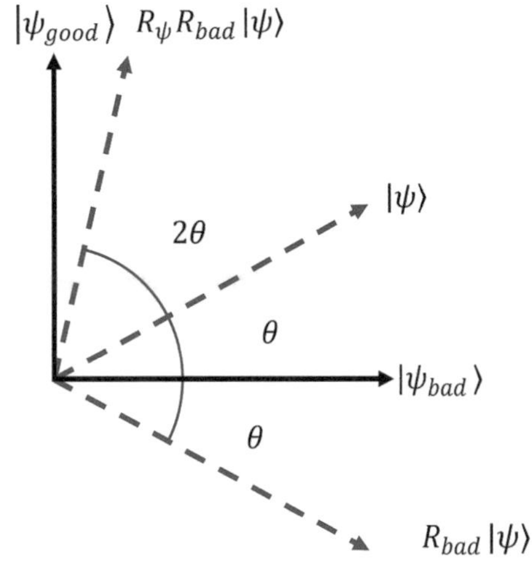

Fig. 24.4 One iteration of amplitude amplification visualized as two reflections

where $\sin\theta = \sqrt{p}$, $\cos\theta = \sqrt{1-p}$. We can see that for $\theta \leq \pi/4$, this will lead to an increase in the probability of measuring $|\psi_{good}\rangle$. We can now formalize this process to obtain the optimum number of applications of $\mathcal{Q}$.

It can be shown [1] that the eigenvectors of $\mathcal{Q}$ are

$$|\psi_{\pm}\rangle = \frac{1}{\sqrt{2}}\left(\frac{1}{\sqrt{p}}|\psi_{good}\rangle \pm \frac{i}{\sqrt{1-p}}|\psi_{bad}\rangle\right)$$

with eigenvalues $\lambda_{\pm} = e^{\pm i2\theta}$.

By expressing $|\psi\rangle$ in the eigenbasis $|\psi_{\pm}\rangle$ and applying $\mathcal{Q}$ j times, we arrive at

$$\begin{aligned}\mathcal{Q}^j|\psi\rangle &= \frac{-i}{\sqrt{2}}\left(e^{(2j+1)i\theta}|\psi_+\rangle - e^{-(2j+1)i\theta}|\psi_-\rangle\right)\\ &= \frac{1}{\sqrt{p}}\sin((2j+1)\theta)|\psi_{good}\rangle + \frac{1}{\sqrt{1-p}}\cos((2j+1)\theta)|\psi_{bad}\rangle\end{aligned}$$

As is apparent, applying $\mathcal{Q}$ too many times will lead to over-rotation, which will subsequently start reducing the probability of measuring the good state. If a is known, choosing $j = \pi/4\theta = O(\sqrt{p})$ leads to

$$\frac{1}{\sqrt{p}}\sin((2j+1)\theta) \geq \frac{1-p}{\sqrt{p}} \geq \max(1-p, p) \geq 1/2 = O(1)$$

which is a quadratic speedup. In many cases p may not be known, but it can be estimated using the amplitude estimation procedure (presented later in this chapter) without losing the quadratic speedup [1]. The overall complexity of amplitude amplification can now be summarized as follows.

Theorem (Amplitude Amplification) *Let $\mathcal{A}$ be an unitary algorithm such that $\mathcal{A}|0\rangle^{\otimes n} = |\psi\rangle = \sqrt{p}|\psi_{good}\rangle + \sqrt{1-p}|\psi_{bad}\rangle$. Given access to unitaries that apply a phase of -1 to the states $|0\rangle^{\otimes n}$ and $|\psi_{good}\rangle$ and choosing $m = \lfloor \pi/4\theta \rfloor$, $\mathcal{Q}^m\mathcal{A}|0\rangle^{\otimes n}$ produces a state with the outcome $|\psi_{good}\rangle$ with at least* $\max(1-a, a) = O(1)$ *probability where* $\sin\theta = \sqrt{p}$ *and* $0 < \theta \leq \pi/2$.

The overall circuit for amplitude amplification is provided in Fig. 24.5.

More advanced algorithms exist for amplitude amplification for specific problems, a notable example being Variable-Time Amplitude Amplification [3], which can provide speedups for algorithms whose average stopping time is smaller than the maximum stopping time by recasting it as a variable-time stopping algorithm. A more efficient version of the Variable-Time Amplitude Amplification subroutine was presented by [4].

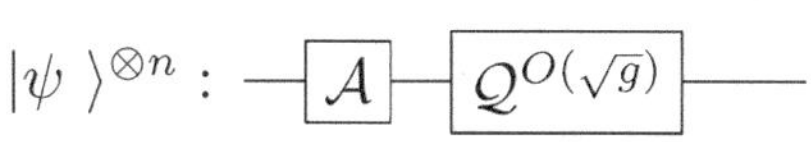

Fig. 24.5 Circuit for amplitude amplification

Grover's search algorithm [5] is a special case of amplitude amplification, which searches for an entry in an unstructured database of N entries using $O\left(\sqrt{N}\right)$ queries to the database compared to $O(N)$ classical queries.

Note that in the formulation above, we have considered $|0\rangle^{\otimes n}$ as the initial state. It may be possible that the input to algorithm $\mathcal{A}$ is a quantum state $|\phi\rangle$ prepared using another quantum algorithm $\mathcal{A}_\phi$. In this case, the operator $\mathcal{S}_0$ (or the reflection R_0) will need to be replaced by an operation marking the state $|\phi\rangle$ (or a Householder reflector about $|\phi\rangle$). In the worst case, this would necessitate multiple invocations of $\mathcal{A}_\phi$.

Quantum Amplitude Estimation

The process of amplitude amplification requires knowledge of the amplitude p corresponding to the probability of measuring a "good" outcome. In many practical quantum algorithms, however, this amplitude is not known in advance. The quantum amplitude estimation (QAE) procedure addresses this by providing a method to estimate p with quadratic improvement in sampling complexity compared to classical techniques.

The setup for amplitude estimation is similar to amplitude amplification [1]. Given a quantum state

$$|\psi\rangle = \sqrt{p}\left|\psi_{good}\right\rangle + \sqrt{1-p}|\psi_{bad}\rangle$$

where

$$\left\langle\psi_{good}|\psi\right\rangle = \sqrt{p}, \qquad \langle\psi_{bad}|\psi\rangle = \sqrt{1-p}$$

the goal is to estimate the unknown success probability p up to a desired precision ϵ.

A naïve approach to estimating $\sqrt{p}$ up to a precision ϵ by directly sampling $|\psi\rangle$ and counting the number of $\left|\psi_{good}\right\rangle$ measurements requires

$$t = \mathcal{O}\left(\frac{\sqrt{p}(1-\sqrt{p})}{\epsilon^2}\right)$$

samples. A quadratic improvement can be made in the number of samples using the QAE procedure.

The quantum amplitude amplification procedure utilizes elements of amplitude amplification and quantum phase estimation. The key idea is to transform the estimation of p into an eigenvalue estimation problem, which is then solved using quantum phase estimation.

From our previous analysis of amplitude amplification, we note that

$$p = \sin^2(\theta)$$

as evident from Fig. 24.4.

Furthermore, we also know that the eigenvalues of $\mathcal{Q}$ are $\lambda_{\pm} = e^{\pm i2\theta}$. Therefore, by applying controlled versions of $\mathcal{Q}$, one may estimate θ as $\tilde{\theta}$, which leads to an estimate of p as $\tilde{p} = \sin^2(\tilde{\theta})$.

If the estimation error in θ satisfies

$$\left|\theta - \tilde{\theta}\right| \leq \epsilon$$

then the error in $\tilde{p}$ satisfies

$$|p - \tilde{p}| \leq 2\epsilon\sqrt{p(1-p)} + \epsilon^2$$

Further analysis of this method leads to the following result.

Theorem (Amplitude Estimation [1]) *For any positive integer k, amplitude amplification produces an estimate $0 < \tilde{p} \leq 1$ s.t.*

$$|p - \tilde{p}| \leq 2\pi k \frac{\sqrt{p(1-p)}}{t} + k^2 \frac{\pi^2}{t^2}$$

with probability at least $8/\pi^2$ when $k = 1$ and with a probability greater than $1 - \frac{1}{2(k-1)}$ for $k \geq 2$ using t evaluations of $\mathcal{A}$.

For further details, we refer to the original work by [1]. The procedure is summarized in the circuit in Fig. 24.6. Once an estimate $\tilde{\theta}$ is obtained, one may estimate $\tilde{p}$ as

$$\tilde{p} = \sin^2(\tilde{\theta}).$$

To obtain $\tilde{p}$ up to a precision ε, the number of samples required is

$$t = \mathcal{O}\left(\frac{\sqrt{p(1-p)}}{\epsilon}\right)$$

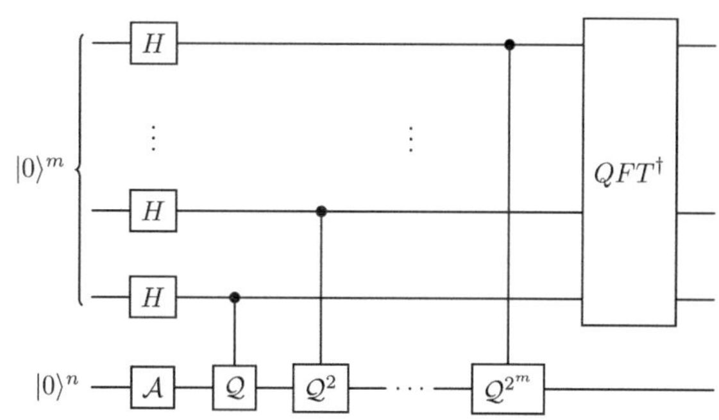

Fig. 24.6 Circuit for quantum amplitude estimation

which is a quadratic improvement over the classical sampling approach.

Similar to quantum phase estimation, the output state is of the form $|y\rangle|\psi\rangle^{\otimes n}$, where $|y\rangle$ encodes the estimated amplitude. This can be extracted as $p \approx \sin^2(\pi\frac{y}{m})$, where m is the number of basis states in the phase estimation register.

This subroutine has numerous applications. For instance, it can be used to estimate the expectation values of unitary operators, as outlined in Chap. 27: Expectation Value Estimation. The quadratic improvement of quantum amplitude estimation over naïve sampling is exploited in the quantum Monte Carlo algorithm to achieve a speedup over classical Monte Carlo, which is presented in the next chapter.

References

1. G. Brassard, P. Høyer, M. Mosca, A. Tapp, Quantum amplitude amplification and estimation. Contemp. Math. **305**; S.J. Lomonaco, H.E. Brandt (eds.), *Providence, Rhode Island: American Mathematical Society* (2002), pp. 53–74. https://doi.org/10.1090/conm/305/05215
2. A.N. Chowdhury, Y. Subasi, R.D. Somma, Improved implementation of reflection operators (2018). https://doi.org/10.48550/ARXIV.1803.02466
3. A. Ambainis, Variable time amplitude amplification and quantum algorithms for linear algebra problems, in *Leibniz International Proceedings in Informatics*, Schloss Dagstuhl - Leibniz-Zentrum für Informatik (2012), pp. 636–647. https://doi.org/10.4230/LIPICS.STACS.2012.636
4. S. Chakraborty, A. Gilyén, S. Jeffery, The power of block-encoded matrix powers: improved regression techniques via faster Hamiltonian simulation, in *Leibniz International Proceedings in Informatics*, Schloss Dagstuhl - Leibniz-Zentrum für Informatik (2019), pp. 1–33. https://doi.org/10.4230/LIPICS.ICALP.2019.33
5. L.K. Grover, A fast quantum mechanical algorithm for database search, in *Proceedings of the Twenty-Eighth Annual ACM Symposium on Theory of Computing—STOC '96* (ACM Press, Philadelphia, Pennsylvania, United States, 1996), pp. 212–219. https://doi.org/10.1145/237814.237866

Quantum Monte Carlo

25

Monte Carlo methods are a class of numerical techniques widely used to approximate integrals and expectation values, particularly in high-dimensional or analytically intractable settings. A prototypical goal is to compute integrals of the form

$$E_f[h(X)] = \int_{\mathcal{X}} f(x)h(x)dx$$

where $E_f[h(X)]$ is the expectation value of a function h under the density function f over a random variable x. The classical Monte Carlo estimate for J samples is

$$\hat{h}_J = \frac{1}{J} \sum_{j=1}^{J} h\left(x^{(j)}\right)$$

where $x^{(j)}$ are independent identically distributed samples over the distribution $x^{(j)} \sim f(x)$. Based on the strong law of large numbers, as $J \to \infty$, $\hat{h}_J \to E_f[h(X)]$. However, to achieve an estimation error $\epsilon = \left|\hat{h}_J - E_f[h(X)]\right|$, the required number of samples J scales as $J = \mathcal{O}\left(\frac{1}{\epsilon^2}\right)$.

Quantum Monte Carlo achieves a quadratic speedup, requiring only $\mathcal{O}\left(\frac{1}{\epsilon}\right)$ samples using quantum amplitude estimation, which we developed in the previous chapter.

Consider 2^n sampling points corresponding to bitstrings x. Assume that an algorithm $\mathcal{A}$ operating on n qubits prepares the state

$$\mathcal{A}|0\rangle^{\otimes n} = \sum_{x=0}^{2^n-1} a_x^2 |x\rangle$$

O. M. Raisuddin and S. De, *Quantum Computing for Engineers*,
https://doi.org/10.1007/978-3-032-03325-3_25

Consider also a function $v(x) : \{0, 1\}^n \rightarrow \mathbb{R}$, mapping the bitstrings x to $v(x)$. We would like to estimate the expectation value

$$E_a[v(\mathcal{A})] = \sum_{x=0}^{2^n-1} |a_x|^2 v(x)$$

In this setting $f(x)$ and $h(x)$ are analogous to $|a_x|^2$ and $v(x)$ respectively. $v(x)$ is accessible through a rotation unitary $\mathcal{R}$ operating on the same n qubits as $\mathcal{A}$, and an ancilla qubit:

$$\mathcal{R}|x\rangle|0\rangle = |x\left[\sqrt{1 - v(x)}|0\rangle + \sqrt{v(x)}|1\rangle\right]$$

Combining these two operations as $\mathcal{F}$

$$\mathcal{F}|0\rangle^{\otimes n+1} = \mathcal{R}(\mathcal{A} \otimes I)|0\rangle^{\otimes n+1}$$
$$= |\chi\rangle = \sum_{x=0}^{2^n-1} a_x|x\left[\sqrt{1 - v(x)}|0\rangle + \sqrt{v(x)}|1\rangle\right]$$

We can see that measuring the probability of measuring the rightmost qubit as $|1\rangle$ for the state $|\chi\rangle$ is

$$p(|\cdot\rangle|1\rangle) = |a_x|^2 v(x) = \mu$$

which is the desired expectation value $\mu = E_a[v(\mathcal{A})]$. Estimating this by repeated sampling leads to a standard error

$$\epsilon = \sqrt{\frac{\mu(1-\mu)}{t}}$$

implying $t = O\left(\frac{\mu(1-\mu)}{\epsilon^2}\right)$ samples for precision ϵ, matching the classical Monte Carlo rate.

To achieve a quadratic improvement, we apply the quantum amplitude estimation (QAE) procedure. We identify the target "good" state as

$$\left|\psi_{good}\right\rangle = |\cdot\rangle|1\rangle$$

so that

$$|\chi\rangle = \sqrt{\mu}\left|\psi_{good}\right\rangle + \sqrt{1-\mu^2}|\psi_{bad}\rangle$$

We will now proceed with constructing the remainder of the ingredients for quantum amplitude estimation.

Consider a unitary which marks the good states $|\cdot\rangle|1\rangle$ with a –ve phase

$$S_{good} = I - 2I \otimes |1\rangle\langle 1|$$

This operator flips the sign of the amplitude on the "good" subspace, and is trivially implemented with a Z gate applied to the rightmost qubit:

$$\begin{aligned}(I \otimes Z)|\chi\rangle &= \sum_{x=0}^{2^n-1} a_x\sqrt{1-v(x)}(I \otimes Z)|x\rangle|0\rangle + \sum_{x=0}^{2^n-1} a_x\sqrt{v(x)}(I \otimes Z)|x\rangle|1\rangle \\ &= \sum_{x=0}^{2^n-1} a_x\sqrt{1-v(x)}|x\rangle|0\rangle - \sum_{x=0}^{2^n-1} a_x\sqrt{v(x)}|x\rangle|1\rangle \\ &= \sqrt{1-\mu^2}|\psi_{bad}\rangle - \sqrt{\mu}\big|\psi_{good}\big\rangle.\end{aligned}$$

Similar to amplitude amplification, we define an operator that reflects across $|\chi\rangle$

$$\mathcal{U} = I - 2|\chi\rangle|\chi\rangle$$

which can be implemented as $\mathcal{U} = \mathcal{F}\mathcal{S}_0\mathcal{F}^\dagger$.

Using these ingredients, we construct the Grover operator

$$\mathcal{Q} = -\mathcal{F}\mathcal{S}_0\mathcal{F}^\dagger S_{good}$$

As we know from the previous chapter, the eigenvalues of $\mathcal{Q}$ are $e^{\pm i2\theta}$ where $\sin\theta = \sqrt{\mu}$. Applying quantum phase estimation to $\mathcal{Q}$ allows us to estimate θ, and thus the desired quantity:

$$\mu = \sin^2(\theta)$$

The result is a quantum algorithm that estimates μ with only

$$t = O\left(\frac{\mu(1-\mu)}{\epsilon}\right)$$

queries, quadratically fewer than the classical sampling approach.

Matrix-Vector Multiplications and Affine Linear Operations

26

Matrix-vector and affine linear operations are foundational in scientific computing. This chapter presents quantum methods to implement these operations using block-encodings, with emphasis on techniques that improve success probability and ancilla overheads.

Matrix-Vector Multiplication Using Block-Encoding

Recall that a matrix A can be encoded in a unitary

$$U_A = \begin{pmatrix} A/\alpha & * \\ * & * \end{pmatrix}$$

where $\alpha \geq \|A\|$ is a normalization constant. Measuring the m ancilla qubits in the state $|0\rangle^{\otimes m}$ yields

$$A/\alpha = \big(\langle 0|^m \otimes I_n\big) U_A \big(|0\rangle^m \otimes I_n\big)$$

A matrix-vector product can be applied as

$$U_A|0\rangle^{\otimes m}|b\rangle = \begin{pmatrix} \frac{A}{\alpha} & * \\ * & * \end{pmatrix} \begin{pmatrix} b \\ 0 \end{pmatrix} = \begin{pmatrix} \frac{Ab}{\alpha} \\ * \end{pmatrix} = \frac{1}{\alpha}|0\rangle^{\otimes m}|Ab\rangle + |\perp\rangle$$

The ancilla qubits can be measured successfully as $|0\rangle^{\otimes m}$ with a probability

$$p\big(|0\rangle^{\otimes m}\big) = \frac{1}{\alpha^2}\| A|b\rangle \|^2$$

O. M. Raisuddin and S. De, *Quantum Computing for Engineers*,
https://doi.org/10.1007/978-3-032-03325-3_26

Sequence of Matrix-Vector Multiplications

One may apply a sequence of matrix-vector products using multiple block encodings. As an example, consider two block encodings U_{A_1} and U_{A_2}, which are $(\alpha_1, m_1, 0)$ and $(\alpha_2, m_2, 0)$ block encodings of A_1 and A_2, respectively as shown in Fig. 26.1. We may prepare the quantum state

$$\frac{1}{\alpha_1\alpha_2}A_2A_1|\psi\rangle$$

by executing the following circuit.

There are a couple of important things to note here. First, the normalization factors accumulate as a product, leading to an exponentially diminishing success probability. Second, the number of ancilla qubits increases as the sum of individual ancilla qubits for each block encoding, i.e., they grow linearly. Two techniques address these issues: the compression gadget reduces ancilla overhead, and uniform singular value amplification (USVA) boosts success probability.

Compression Gadget

The compression gadget [1, 2] is a transformation of a quantum circuit applying a sequence of matrix-vector multiplications into an equivalent one with fewer ancilla qubits. Consider a sequence of $(\alpha_i, m_i, 0)$ block encodings U_{A_i} for matrices A_i, which are applied to a quantum state $|\psi\rangle$ to get

$$\prod_{i=L}^{1}\frac{1}{\alpha_i}A_i|\psi\rangle$$

Denoting $m_{\max} = \max_i m_i$ as the largest number of ancilla needed for each individual block encoding and defining $\lambda = \log_2 L + 1$, the unitary operation ADD is defined as

$$ADD|i\rangle = \left|\operatorname*{mod}_{\lambda}(i+1)\right\rangle$$

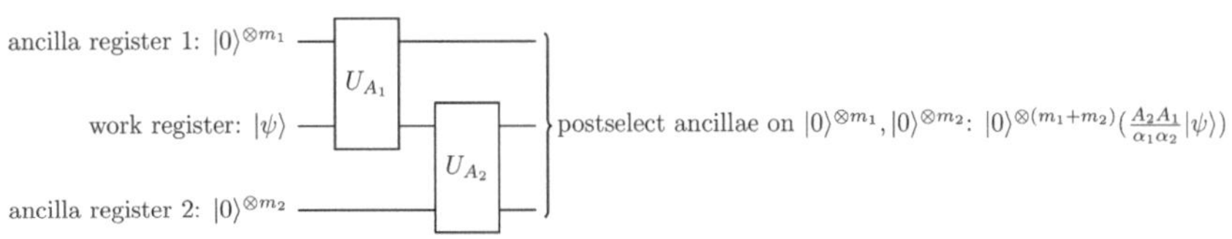

Fig. 26.1 Circuit for applying a sequence of 2 matrix-vector multiplications with two separate block encodings

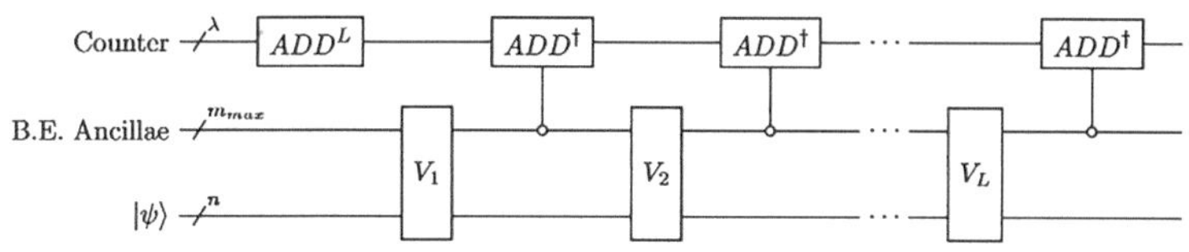

Fig. 26.2 Quantum circuit for a compression gadget

The compression gadget may be constructed as shown in Fig. 26.2.

The block encodings reuse the $m_{\max}$ qubits, at the expense of λ additional ancilla qubits for a "counter."

Uniform Singular Value Amplification

We now address the issue of exponentially decaying success probabilities for a sequence of matrix multiplications using the block-encoding method. This is achieved by using a block encoding U_i to create another block-encoding with a higher success probability (smaller normalization factor). These boosted block encodings are then used to compute the matrix-vector products.

The uniform singular value amplification (USVA) [2, 3] method boosts the normalization factor of each block encoding toward the largest possible value, i.e., the intrinsic normalization factor $\alpha_{min} = \|A\|$. This is achieved by fitting an odd polynomial to a linear function

$$P(x) = \frac{1-\delta}{\alpha_i} x$$

over the interval $x \in [0, \alpha_i]$ using quantum signal processing, as shown in Fig. 26.3. This is done individually for each block U_{A_i}.

Computing the matrix polynomial $P(A_i)$ then yields a new block-encoding of A_i with a smaller normalization constant. More specifically, it is a $\left(\frac{\|A_i\|}{(1-\delta)}, m_i + 1, \tilde{\epsilon}_i \|A_i\|\right)$ block-encoding of A_i, and it requires

Fig. 26.3 Visualization of the operation of uniform singular value amplification

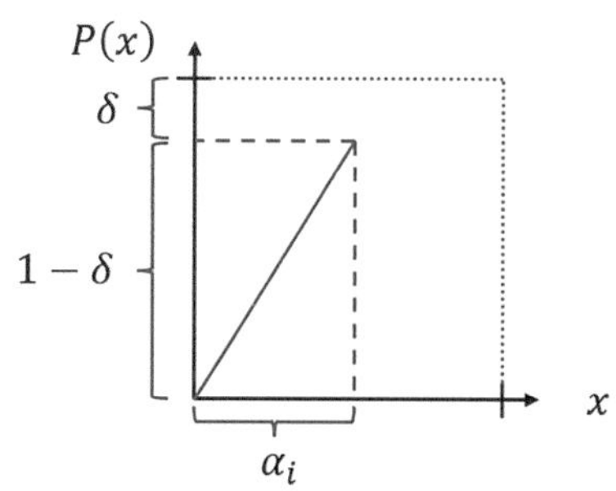

$O\left(\frac{\alpha_i}{\delta\|A_i\|}\log\left(\frac{\alpha_i}{\|A_i\|\tilde{\epsilon}_i}\right)\right)$ accesses to (controlled versions of) U_{A_i} and its adjoint (QSP sequence has a phase angle sequence of length, or $P(x)$ is a polynomial of degree $O\left(\frac{\alpha_i}{\delta\|A_i\|}\log\left(\frac{\alpha_i}{\|A_i\|\tilde{\epsilon}_i}\right)\right)$). The error $\tilde{\epsilon}_i$ in the new block encoding arises from the imperfect polynomial fit through QSP, and may be reduced exponentially by increasing the number of phase angles in the QSP sequence.

USVA Lemma [2]: Let U_A be a $(\alpha, m, 0)$ block encoding of A. A $\left(\frac{\|A\|}{1-\delta}, m+1, \epsilon\|A\|\right)$ block encoding $\tilde{U}_A$ of A can be constructed using $\mathcal{O}\left(\frac{\alpha}{\delta\|A\|}\log\left(\frac{\alpha}{\|A\|\epsilon}\right)\right)$ applications of controlled-U_A and $U_A^{\dagger}$.

We may now use these boosted block encodings, denoted by $\tilde{U}_{A_i}$, to compute the matrix-vector products (using a compression gadget to reduce ancilla overheads). We state the result of [2] in the following theorem, which considers the more general case where U_{A_i} is a $(\alpha_i, m_i, \epsilon_i\|A_i\|)$ block encoding of A_i.

Theorem ([2]) *Given $(\alpha_i, m_i, \epsilon_i\|A_i\|)$ block encodings U_{A_i} of A_i for $1 \leq i \leq L$ where $\Sigma_i \epsilon_i \leq \frac{1}{2}$. For any $0 \leq \epsilon' \leq \frac{1}{2L}$ a (α', m', ϵ'), block encoding of $\Pi_{i=L}^{1} A_i$ can be constructed where*

$$\frac{\Pi_{i=1}^{L}\|A_i\|}{2(1-\delta)^L} \leq \alpha' \leq \frac{e^{1/2}\,\Pi_{i=1}^{L}\|A_i\|}{2(1-\delta)^L}$$

$$m' = m_{\max} + \log_2 L + 2$$

$$\epsilon' = e^{\frac{1}{2}}\left(L\epsilon' + \sum_{i=1}^{L}\epsilon_l\right)\prod_{i=1}^{L}\|A_i\|$$

using $O\left(\frac{\alpha_i}{\delta\|A_i\|}\log\left(\frac{\alpha_i}{\|A_i\|\epsilon'}\right)\right)$ (controlled) applications of each U_{A_i} and its adjoint.

In summary, the USVA procedure requires one additional ancilla qubit, leading to $m_{\max}+1$ ancillae for each boosted block encoding $\tilde{U}_{A_i}$. Using a compression gadget to apply the matrix products leads to a total of $m_{\max} + \log_2 L + 2$ ancilla qubits using total of $O\left(\Sigma_{i=1}^{L}\frac{\alpha_i}{\delta\|A_i\|}\log\left(\frac{\alpha_i}{\|A_i\|\epsilon'}\right)\right)$ applications of block encodings of U_{A_i}. The success probability of the final product is

$$p\left(|0\rangle^{\otimes(m_{\max}+\log_2 L+2)}\right) = \left[\frac{e^{\frac{1}{2}}\prod_{i=1}^{L}\|A_i\|}{2(1-\delta)^L}\left(\prod_{i=L}^{1}A_i|\psi\rangle\right)\right]^2$$

with an error of

$$\epsilon' = e^{\frac{1}{2}}\left(L\epsilon' + \sum_{i=1}^{L}\epsilon_l\right)\prod_{i=1}^{L}\|A_i\|.$$

Affine Linear Operations

So far, we have developed methods to apply matrix-vector products (linear operations) of the form

$$\prod_{i=L}^{1} A_i x$$

We now turn our attention to the more general sequences of affine linear operations of the general form:

$$x^{(i+1)} = Ax^{(i)} + b^{(i)}$$

We present two methods for encoding affine linear operations on quantum computers: solving a block-linear system of equations and performing block-matrix multiplications. As an example, we will consider the following sequence of affine linear operations to demonstrate the two techniques

$$x_1 = A_0 x_0 + b_0$$

$$x_2 = A_1 x_1 + b_1$$

After presenting these two methods, we explain a post-processing step to extract the desired final vector and methods to boost the success probability of this post-processing step.

Block-Linear System of Equations:

$$\begin{pmatrix} I & & \\ -A_0 & I & \\ & -A_1 & I \end{pmatrix} \begin{pmatrix} x_0 \\ x_1 \\ x_2 \end{pmatrix} = \begin{pmatrix} x_0 \\ b_0 \\ b_1 \end{pmatrix}$$

Solving this linear system of equations yields a column of vectors

$$\begin{pmatrix} x_0 \\ x_1 \\ x_2 \end{pmatrix} = \begin{pmatrix} x_0 \\ A_0 x_0 + b_0 \\ A_1 x_1 + b_1 \end{pmatrix}$$

This block-matrix system of equations may be solved using direct quantum linear system algorithms (QLSA), which are introduced in Lecture 30 Quantum Linear System Algorithms: Direct Methods. Efficient application of quantum linear system algorithms requires the linear system to be well-conditioned. For the case $\|A_i\| \leq 1$, the linear system is well-conditioned, i.e., for $1 \leq i \leq l$, $\kappa_M, = 2l$ where M is the block-linear system. Note that the QLSA will introduce a solution error.

Block-Matrix Multiplication

The following sequence of matrix multiplications $M_2 M_1 b = x$ yields the same result

$$\begin{pmatrix} I & & \\ & I & \\ & A_1 & I \end{pmatrix}\begin{pmatrix} I & & \\ A_0 & I & \\ & & I \end{pmatrix}\begin{pmatrix} x_0 \\ b_0 \\ b_1 \end{pmatrix}$$

$$= \begin{pmatrix} I & & \\ & I & \\ & A_1 & I \end{pmatrix}\begin{pmatrix} x_0 \\ A_0 x_0 + b_0 \\ b_1 \end{pmatrix}$$

$$= \begin{pmatrix} I & & \\ & I & \\ & A_1 & I \end{pmatrix}\begin{pmatrix} x_0 \\ x_1 \\ b_1 \end{pmatrix} = \begin{pmatrix} x_0 \\ x_1 \\ A_1 x_1 + b_1 \end{pmatrix} = \begin{pmatrix} x_0 \\ x_1 \\ x_2 \end{pmatrix}$$

$$U_{M_1} = \begin{pmatrix} M_1/\alpha_1 & * \\ * & * \end{pmatrix}; \quad U_{M_2} = \begin{pmatrix} M_2/\alpha_2 & * \\ * & * \end{pmatrix}$$

The procedures presented above for a sequence of matrix multiplications may then be used to implement this operation.

Post-processing and Boosting Success Probabilities

The two methods presented earlier prepare a quantum state encoding the entire sequence of vectors as

$$|x\rangle = \sum_{i=0}^{l} |i\rangle|x_i\rangle = \begin{pmatrix} |x_0\rangle \\ |x_1\rangle \\ \vdots \\ |x_2\rangle \end{pmatrix}$$

The first register $|i\rangle$ is often referred to as an index register or a Feynman–Kitaev clock, and the second register $|x_i\rangle$ is the work register. To prepare the desired final state $|x_i\rangle$, the index register needs to be measured. Measuring the index register in the state $|l\rangle$ indicates that the work register is in the state $|x_i\rangle$. The probability of a successful measurement is

$$p(|l\rangle) = \frac{\| |x_l\rangle \|^2}{\| |x\rangle \|^2}$$

which diminishes as the number of steps l increases. To circumvent this issue, a sequence of p "copy" steps can be appended to the sequence of operations:

$$x_{l+i} = x_{l+i-1} \;\forall\, i \in \left[1, p\right]$$

which, in the case of the block-linear system of equations approach, can be appended to the existing matrix as

$$\begin{pmatrix} I & & \\ -A_0 & I & \\ & -A_1 & I \end{pmatrix}\begin{pmatrix} x_0 \\ x_1 \\ x_2 \end{pmatrix} = \begin{pmatrix} x_0 \\ b_0 \\ b_1 \end{pmatrix} \to \begin{pmatrix} I & & & & \\ -A_0 & I & & & \\ & -A_1 & I & & \\ & & -I & I & \\ & & & -I & I \end{pmatrix}\begin{pmatrix} x_0 \\ x_1 \\ x_2 \\ x_2 \\ x_2 \end{pmatrix} = \begin{pmatrix} x_0 \\ b_0 \\ b_1 \\ 0 \\ 0 \end{pmatrix}$$

Note that this does not violate the no-cloning theorem. The boosted success probability is

$$p_{succ} = \sum_{i=l}^{l+p} p(|i\rangle) = \sum_{i=0}^{p} \frac{\| |x_{l+i}\rangle \|^2}{\| |x\rangle \|^2} = (p+1)\frac{\| |x_l\rangle \|^2}{\| |x\rangle \|^2}$$

The success probability can be further boosted by a quadratic factor using amplitude amplification.

As an example, for the block-linear system approach, appending these copy steps yields the block matrix:

$$\begin{pmatrix} I & & \\ -A_0 & I & \\ & -A_1 & I \end{pmatrix}\begin{pmatrix} x_0 \\ x_1 \\ x_2 \end{pmatrix} = \begin{pmatrix} x_0 \\ b_0 \\ b_1 \end{pmatrix}$$

$$\begin{pmatrix} I & & & & & & \\ * & I & & & & & \\ & \ddots & \ddots & & & & \\ & & * & I & & & \\ & & & -I & I & & \\ & & & & \ddots & \ddots & \\ & & & & & -I & I \end{pmatrix}\begin{pmatrix} x_0 \\ * \\ \vdots \\ x_l \\ x_l \\ x_l \\ x_l \end{pmatrix} = \begin{pmatrix} x_0 \\ * \\ \vdots \\ b_{l-1} \\ 0 \\ \vdots \\ 0 \end{pmatrix}$$

Note that the measurement operation will also lead to renormalization due to the Born rule, and any error ϵ in $|x\rangle$ will be amplified in $|x_l\rangle$ by a factor of $\frac{\| |x\rangle \|^2}{\| |x_l\rangle \|^2}$.

The techniques presented in this chapter are used extensively to develop quantum algorithms for differential equations and iterative quantum linear system algorithms.

References

1. G.H. Low, N. Wiebe, Hamiltonian simulation in the interaction picture (2019). arXiv:1805.00675. https://doi.org/10.48550/arXiv.1805.00675

2. D. Fang, L. Lin, Y. Tong, Time-marching based quantum solvers for time-dependent linear differential equations. Quantum **7**, 955 (2023). https://doi.org/10.22331/q-2023-03-20-955
3. A. Gilyén, Y. Su, G.H. Low, N. Wiebe, Quantum singular value transformation and beyond: exponential improvements for quantum matrix arithmetics, in *Proceedings of the 51st Annual ACM SIGACT Symposium on Theory of Computing* (ACM, Phoenix, AZ, USA, 2019), pp. 193–204. https://doi.org/10.1145/3313276.3316366

Part VI
Quantum Algorithms

This part provides an overview of foundational algorithms developed for quantum computers, with a focus on their application to scientific and engineering computation. Developing quantum algorithms necessitates a fundamental rethinking due to key differences between classical computing models, such as the Turing machine, and the gate-based quantum computing model. While classical algorithms have benefited from decades of refinement, specialized libraries (such as BLAS for linear algebra), and dedicated hardware accelerators, quantum computing remains in an early stage. Progress is shaped by both hardware limitations, such as error rates and coherence times, and the development of new algorithmic paradigms.

The field of quantum algorithms is broad and includes celebrated results such as Shor's factoring and Grover's search; however, we deliberately leave out several well-known algorithms that, while foundational, have limited direct impact on the kinds of large-scale numerical and simulation problems faced by computational engineers. Instead, we focus on core algorithmic techniques—such as expectation value estimation, Hamiltonian simulation, quantum linear system solvers, differential equation solvers, and variational methods—that form the backbone of quantum scientific computing. These algorithms are chosen for both their conceptual significance and their practical relevance to problems where quantum computers are expected to offer genuine advantages over classical approaches. Our goal is to provide readers with a toolkit of quantum algorithms that are most likely to influence the development of computational science as quantum hardware matures, while being candid about current limitations and open challenges.

Chapter 27, "Expectation Value Estimation", presents a variety of quantum techniques for extracting physical observables and statistical quantities from quantum states, providing the foundation for scientific and engineering applications of quantum algorithms.

Chapter 28, "Hamiltonian Simulation Techniques", introduces the central challenge of simulating quantum time evolution, reviewing several quantum approaches for approximating the dynamics of quantum systems relevant to chemistry, materials, and physics.

Chapter 29, "Eigenvalue Problems", surveys quantum algorithms for computing eigenvalues and eigenvectors of matrices, with applications to spectral analysis, principal component methods, and quantum chemistry.

Chapter 30, "Quantum Linear System Algorithms: Direct Methods", covers direct quantum algorithms for solving linear systems of equations, such as the

Harrow–Hassidim–Lloyd (HHL) algorithm, and discusses conditions for efficient implementation.

Chapter 31, "Quantum Linear System Algorithms: Iterative Methods", presents iterative quantum algorithms designed for large or structured linear systems, exploring their potential for exponential speedup and discussing issues such as preconditioning.

Chapter 32, "Quantum Ordinary Differential Equation Algorithms: Block-Matrix Algorithms", describes methods for mapping systems of ordinary differential equations (ODEs) to block-linear systems solvable by quantum algorithms, addressing both homogeneous and inhomogeneous problems.

Chapter 33, "Quantum Ordinary Differential Equation Algorithms: Time-Marching Algorithms", introduces quantum algorithms that simulate the time evolution of ordinary differential equations (ODE) using time-marching schemes.

Chapter 34, "Quantum Partial Differential Equation Algorithms", explores quantum algorithms for partial differential equations (PDEs), focusing on strategies for discretization, reduction to ODE systems, and direct quantum solution techniques for high-dimensional problems in engineering and the sciences.

Chapter 35, "Variational Algorithms: Theory", develops the mathematical and algorithmic framework for variational quantum algorithms, which employ hybrid quantum-classical optimization to solve problems in simulation, optimization, and machine learning.

Chapter 36, "Notable Variational Algorithms", surveys prominent variational quantum algorithms—including the Variational Quantum Eigensolver (VQE), Quantum Approximate Optimization Algorithm (QAOA), and Variational Quantum Linear Solver (VQLS)—and discusses their applications in quantum chemistry, combinatorial optimization, and linear algebra.

Chapters 27–34 are ordered to build upon one another, with earlier chapters covering more mature techniques and later chapters highlighting emerging methods and recent progress in quantum algorithm development.

27 Expectation Value Estimation

Estimating an expectation value of a quantum state $|\psi\rangle$ for an operator H is defined as the task of estimating

$$\langle H_{\psi} \rangle = \langle \psi | H | \psi \rangle$$

where H is either a Hermitian or unitary operator. For Hermitian operators, this is often referred to as "measuring the observable H."

Note: The symbol H is also used to denote the Hadamard gate in quantum computing. Although this notation is standard in the literature, we alert the reader to this potential source of confusion and clarify usage whenever necessary.

There are several standard techniques for expectation value estimation. We summarize three widely used procedures in Table 27.1.

Pauli Diagonalization

Pauli Diagonalization is the most widely used method in NISQ algorithms. The procedure does not require any ancilla qubits.

We begin by considering an arbitrary Hermitian matrix H. Since H is Hermitian, it can be diagonalized as $H = V^{\dagger} D V$, allowing us to rewrite the expectation value as

$$\langle H \rangle_{\psi} = \langle \psi | H | \psi \rangle = \langle \psi | V^{\dagger} D V | \psi \rangle = \langle \psi' | D | \psi' \rangle = \sum_{i} D_{ii} \left| \langle i| \, ' \rangle \right|^2$$

where $|\psi'\rangle = V|\psi\rangle$. If we have access to a procedure to apply V, we can prepare $|\psi'\rangle$, sample measurement outcomes in the diagonal basis of D, and estimate the

O. M. Raisuddin and S. De, *Quantum Computing for Engineers*,
https://doi.org/10.1007/978-3-032-03325-3_27

Table 27.1 Summary of various techniques for expectation value estimation

Procedure	Sample complexity	Ancillae
Pauli diagonalization	$O(1/\epsilon^2)$	None
Hadamard test	$O(1/\epsilon^2)$	1
Phase estimation	$O(1/\epsilon)$	$O(\log 1/\epsilon)$

probability distribution

$$p(i) = \langle \psi' | M_i^\dagger M_i | \psi' \rangle$$

by sampling bitstrings from $|\psi'\rangle$ to get an estimate $\tilde{p}(i)$ of $p(i)$. The expectation value can then be approximated as

$$\langle \psi | H | \psi \rangle \approx \sum_i D_{ii} \tilde{p}(i)$$

However, in practice, this is very unlikely to be feasible in general since V and D_{ii} may not be computable efficiently, or an efficient procedure to implement V on a quantum computer may not be available.

Fortunately, in problems arising in quantum simulation, H can be expressed as a linear combination $H = \sum_j \alpha_j H_j$, where $\alpha_j \in \mathbb{C}, H_j \in \mathbb{C}^{2^n \times 2^n}$ of simpler operators H_j in the form of strings of Pauli gates or operators, often referred to as Pauli strings. An example of a Pauli string H_j is

$$H_j = I \otimes X \otimes Y \otimes I \otimes Z$$

These sums are also typically sparse, i.e., contain relatively few terms compared to the dimension of the problem. Replacing H with its additive decomposition $\sum_j \alpha_j H_j$, we get

$$\langle H_\psi \rangle = \langle \psi | H | \psi \rangle = \langle \psi | \sum_j \alpha_j H_j | \psi \rangle = \sum_j \alpha_j \langle \psi | H_j | \psi \rangle$$

where

$$H_{ex} = \sigma_a \otimes \sigma_b \otimes \ldots \otimes \sigma_n = \bigotimes_{\substack{j \\ k}} \sigma_k$$

such that $\sigma_k \in \{I, X, Y, Z\}$ as $\sigma_0, \sigma_1, \sigma_2, \sigma_3 = I, X, Y, Z$ respectively. Using the eigendecompositions of individual Pauli gates, $\sigma_k = V_k^\dagger D_k V_k$, provided in Table 8, this may be transformed into

$$\begin{aligned}\langle H\rangle_\psi &= \sum_j \alpha_j \langle\psi|H_j|\psi\rangle = \sum_j \alpha_j \langle\psi|\bigotimes_j \sigma_k|\psi\rangle \\ &= \sum_j \alpha_j \langle\psi|\bigotimes_j V_k^\dagger D_k V_k|\psi\rangle = \sum_j \alpha_j \left\langle\psi_j'\right|\bigotimes_j D_k\left|\psi_j'\right\rangle \\ &= \sum_j \alpha_j \left\langle\psi_j'\right|D_j\left|\psi_j'\right\rangle = \sum_i\sum_j \alpha_j D_{j_{ii}} \left|\left\langle i|{}_j'\right\rangle\right|^2\end{aligned}$$

In the simplest setting, given $|\psi\rangle$, the procedure consists of

(i) Preparing the states $\left|\psi_j'\right\rangle$ by applying the (known) operators $\bigotimes_{\substack{j\\k}} V_k$ on the state $|\psi\rangle$.
(ii) Sampling bitstrings in the computational basis.
(iii) Computing the double sums over i and j.

To compute the terms $D_{j_{ii}}$ we first define a bitstring $\mathscr{b}_j$ whose bits are set to 1 for each operator $\sigma_k \neq I$ in H_j and 0 otherwise, from left to right in the same order as the Kronecker product order of H_j. Using the same example Pauli string H_{ex} above, we obtain

$$\mathscr{b}_{ex} = 01101$$

$D_{k_{ii}}$ can be computed using the parity (number of 1's), $\mathscr{p}_{lj}$, of the bitstring obtained from a bitwise AND operation (denoted as &) on the measured bitstring i and $\mathscr{b}_j$, i.e.,

$$\mathscr{p}_{lj} = \text{parity}(i \,\&\, b_j)$$

$$D_{k_{ii}} = \{ \begin{matrix} 1 & \mathscr{p}_{lj} \in even \\ -1 & \mathscr{p}_{lj} \in odd \end{matrix}$$

Using the same example, we consider the case where we sample a bitstring $l = \text{bin}(21) = 10101$ from $\left|\psi'\right\rangle$. This leads to $\mathscr{p}_{l,ex} = \text{parity}(10101 \,\&\, 01101) = \text{parity}(00101) = 2$, and therefore $D_{ex_{ll}} = 1$.

We now consider another example of a quantum state $|\phi\rangle = \alpha|00\rangle + \beta|01\rangle + \gamma|10\rangle + \delta|11\rangle$ and an operator

$$H = H_1 + H_2$$

where

$$H_1 = aX \otimes Y$$

$$H_2 = bZ \otimes I$$

Using the eigendecompositions of Pauli gates, we obtain

$$H_1 = aHZH \otimes SHZHS^{\dagger}$$

$$H_2 = bZ \otimes I$$

Therefore, the states $\left|\phi_1'\right\rangle = H \otimes HS^{\dagger}|\phi\rangle$ and $\left|\phi_2'\right\rangle = Z \otimes I|\phi\rangle$ are prepared on a quantum computer and sampled.

Suppose that a total of 1024 samples are obtained (512 each for $\left|\phi_1'\right\rangle$ and $\left|\phi_2'\right\rangle$ with the following statistics:

Measured string i	# of samples for $\left\|\phi_1'\right\rangle$	# of samples for $\left\|\phi_2'\right\rangle$
$\|00\rangle$	105	76
$\|01\rangle$	253	97
$\|10\rangle$	82	312
$\|11\rangle$	72	27

The bitstrings $\mathscr{b}_1$ and $\mathscr{b}_2$ corresponding to H_1 and H_2 are then computed by setting bits to 1 in each position where a Pauli gate is applied:

$$\mathscr{b}_1 = 11$$
$$\mathscr{b}_2 = 10$$

We may compute the entries $D_{k_{ii}}$ as

Measured string i	$i \,\&\, \mathscr{b}_1$	$i \,\&\, \mathscr{b}_2$	p_{i1}	p_{i2}
$\|00\rangle$	00	00	1	1
$\|01\rangle$	01	00	−1	1
$\|10\rangle$	10	10	−1	−1
$\|11\rangle$	11	10	1	−1

The expectation value may then be computed as

$$\langle H\rangle_{\phi} = a\left(\frac{105}{512} + (-1)\frac{253}{512} + (-1)\frac{82}{512} + \frac{72}{512}\right)$$

$$+b\left(\frac{76}{512}+\frac{97}{512}+(-1)\frac{312}{512}+(-1)\frac{27}{512}\right)$$

Typically, quantum computing libraries automatically perform these computations for sums of Pauli strings. The following code compares the Qiskit implementation of expectation value computations with a custom implementation. The code first samples quantum states $|\psi'\rangle$ for each Pauli string and post-processes the bitstrings. These values are then compared with Qiskit's built-in procedure for computing expectation values exactly:

```
#!/usr/bin/python3

import numpy as np
import random
from qiskit import QuantumCircuit, QuantumRegister, ClassicalReg-
ister
# from qiskit_aer import UnitarySimulator, StatevectorSimulator
# from qiskit.primitives import Estimator, Sampler
from qiskit_aer.primitives import SamplerV2 as Sampler, Estima-
torV2 as Estimator
from qiskit.quantum_info import SparsePauliOp
from qiskit.circuit.library import SGate
from qiskit.circuit.random import random_circuit

# Create a random quantum circuit
seed = 12345
n = 4
circ = random_circuit(n, n*2, seed=seed)

# Define some random observables as Pauli operators
num_paulis = 10
ops = []
pauli_pool = ['I','X','Y','Z']
 for _i in range(num_paulis):
     pauli_string = ''
     for _j in range(n):
                         pauli_string = pauli_string + pauli_
pool[random.randint(0,3)]
     ops.append(SparsePauliOp(pauli_string,random.random()))
# Combine into one operator
H = sum(ops)

# Create circuits corresponding to Pauli string diagonalizations:
# Applying the appropriate eigenvectors to each qubit for the cor-
responding Pauli gate in the Pauli string
```

```
obs_circs = []
for pauli_string in H.to_list():
    pauli_circ = QuantumCircuit(n)
    for index, pauli in enumerate(pauli_string[0]):
        if pauli=='X':
            pauli_circ.h(n-1-index)
        elif pauli=='Y':
            pauli_circ.append(SGate().inverse(),[n-1-index])
            pauli_circ.h(n-1-index)
        obs_circs.append((pauli_circ,pauli_string))

# Sample these circuits and package into a tuple with
# quasiprobabilities, Pauli strings, and coefficients of the Pauli
strings
sampler = Sampler()
qp_ps_coeff = []
for obs_circ in obs_circs:
    newcirc = circ.compose(obs_circ[0])
    newcirc.measure_all()
    job = sampler.run([newcirc.decompose()],shots=10_000_000)
    result = job.result()[0].data.meas.get_counts()
    # Get the quasi-probabilities from the result
     # The result is a dictionary with bitstrings as keys and their
counts as values
    total_counts = sum(result.values())
    for key in result:
    fresult[key] = result[key] / total_counts
    fq_p = result
    fqp_ps_coeff.append((q_p,obs_circ[1][0],obs_circ[1][1]))

# Estimate the expectation values
evs = []
for _i in qp_ps_coeff:
    # Unpack tuple
    qp = _i[0]
    ps = _i[1]
    coeff = _i[2].real

# Create bitstring from Pauli string
# Any qubit with a Pauli applied will be marked as 1 in the bitstring
pauli_bs = ''
for _j in ps:
    if (_j == 'I'):
        pauli_bs = pauli_bs + '0'
    else:
```

```
        pauli_bs = pauli_bs + '1'
    print(f'Pauli string: {ps} Bitstring: {pauli_bs}\n')
    print(f'Quasi-probs: {qp} \n')

    # Expectation value is computed as a sum over all the
    # quasiprobabilities of the sampled bitstrings
    ev = 0
    for index_bs in qp:
        # Get quasiprobability
        _k = qp[index_bs]
        print('Index BS: ' + index_bs)
        print('Pauli BS: ' + pauli_bs)
        # Perform bitwise AND with these two
        # This will tell us how many -ve signs are being picked up
            bs_mask_ps = f'{{0:0{n}b}}'.format(int(index_bs,2) &
int(pauli_bs,2))
        print('Bitwise AND: ' + bs_mask_ps)
        # Compute parity of this final bitstring
        # This will give the overall number of -ve signs
        parity = 1
        for _l in bs_mask_ps:
          if _l=='1':
            parity = parity*-1
        print(f'Parity: {parity}')
        print()
        # Compute the contribution of the sampled bitstring + its
quasiprobability
        # towards the expectation value
        ev = ev + parity * _k
        evs.append(coeff * ev)
        print()
        print()

# Print out expectation values of each Pauli string
print(np.array(evs))

# Estimate observables using Qiskit's Estimator method and compare
estimator = Estimator()
job = estimator.run([(circ.decompose(),o) for o in H])
qiskit_estimator_result = [_result.data.evs for _result in
job.result()]
print(qiskit_estimator_result)

# Compute total error for each expectation value
```

```
print(f'Total error for each expectation value:
{np.abs(np.sum(evs) - np.sum(qiskit_estimator_result))}')
```

The final line of output is the difference between the two methods:

```
Total error for each expectation value: 0.00017310799751490968
```

which is within the expected bounds for sampling error (scales as $O(1/\epsilon^2)$).

It may be possible for various Pauli strings to commute. As an example, consider the Pauli strings

$$H_1 = \alpha_1 I \otimes Z \otimes X, H_2 = \alpha_2 X \otimes Z \otimes I$$

With the mixed-product property of Kronecker products in mind, we note that each Pauli matrix in the first Pauli string commutes with the corresponding Pauli matrix in the second Pauli string, i.e.,

$$[I, X] = [Z, Z] = [X, I] = 0$$

implying that these two Pauli strings commute. Therefore, they must share the same eigenvectors V_1, V_2. Using this fact, instead of preparing and sampling $|\psi_1'\rangle = V_1|\psi\rangle, |\psi_2'\rangle = V_2|\psi\rangle$ individually for these two Pauli strings, one may simply sample $|\psi_{1,2}'\rangle = V_1|\psi\rangle = V_2|\psi\rangle$ and then compute $\sum_i (\alpha_1 D_{1ii} + \alpha_2 D_{2ii}) \left|\langle i|'_{1,2}\rangle\right|^2$ for better sampling efficiency. This is more apparent by diagonalizing H_1 and H_2:

$$H_1 = \alpha_1 I \otimes Z \otimes HZH = \alpha_1 HH \otimes Z \otimes HZH$$

$$H_2 = \alpha_2 HZH \otimes Z \otimes I = \alpha_2 HZH \otimes Z \otimes HH$$

$$H_1 + H_2 = H(\alpha_1 I + \alpha_2 Z)H \otimes (\alpha_1 Z + \alpha_2 Z) \otimes (\alpha_1 Z + \alpha_2 I)$$

Note that for Pauli strings to commute, each Pauli matrix in the first string does not need to commute with the corresponding Pauli matrix in the second string. As an example, consider

$$H_3 = \alpha_3 Z \otimes Y \otimes Z, \quad H_4 = \alpha_4 X \otimes Z \otimes I$$

Not every individual Pauli matrix in the first stringcommutes with the second string, e.g., $[Z, X] \neq 0$. Regardless, H_3 and H_4 do commute:

$$[Z \otimes Y \otimes Z, X \otimes Z \otimes I] = [Z, X] \otimes [Y, Z] \otimes [Z, I] = [Z, X] \otimes [Y, Z] \otimes 0 = 0$$

Efficiently partitioning Pauli strings into groups of commuting strings is an active area of research. A recent implementation and review can be found in [1].

There are other methods in the literature to estimate expectation values using various decompositions and measurement bases [2]. However, the Pauli string decomposition is used most commonly due to its simplicity and generality.

Hadamard Test

We now introduce the Hadamard test for estimating the expectation value $\langle\psi|U|\psi\rangle$ of a unitary $U \in \mathbb{C}^{2^n \times 2^n}$. Since U is unitary rather than Hermitian, $\langle\psi|U|\psi\rangle \in \mathbb{C}$, i.e., the expectation value is no longer guaranteed to be real.

The real and imaginary parts of the expectation value may be estimated using the following circuits shown in Fig. 27.1.

To see how this works, consider a state $|\psi\rangle$. Adding the ancilla qubit, we get

$$|0\rangle|\psi\rangle$$

Applying the first Hadamard gate, we obtain

$$(H \otimes I)|0\rangle|\psi\rangle = \frac{1}{\sqrt{2}}(|0\rangle + |1\rangle)|\psi\rangle$$

Subsequently, after applying the controlled version of U we get

$$(cU)\frac{1}{\sqrt{2}}(|0\rangle + |1\rangle)|\psi\rangle = (|0\rangle\langle 0|I + |1\rangle\langle 1|U)\frac{1}{\sqrt{2}}(|0\rangle + |1\rangle)|\psi\rangle$$
$$= \frac{1}{\sqrt{2}}(|0\rangle|\psi\rangle + |1\rangle U|\psi\rangle)$$

and finally, applying the final Hadamard gate yields

$$\frac{1}{\sqrt{2}}(H \otimes I)(|0\rangle|\psi\rangle + |1\rangle U|\psi\rangle) = \frac{1}{2}((|0\rangle + |1\rangle|\psi\rangle) + (|0\rangle - |1\rangle)U|\psi\rangle)$$
$$|\phi\rangle = \frac{1}{2}|0\rangle(|\psi\rangle + U|\psi\rangle) + \frac{1}{2}|1\rangle(|\psi\rangle - U|\psi\rangle)$$

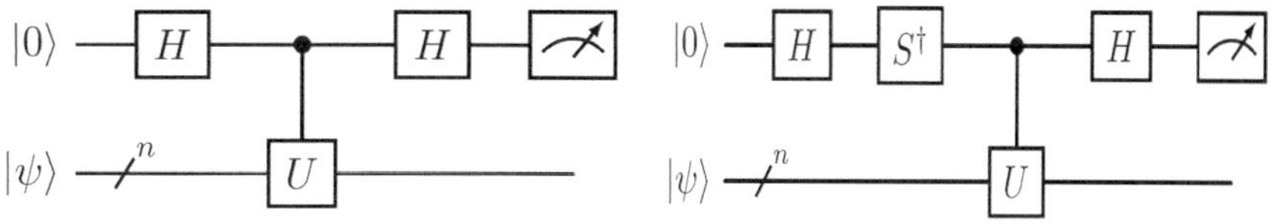

Fig. 27.1 Hadamard test circuits. Left: real part; right: imaginary part

Now we analyze the probability of measuring the ancilla qubit in the state $|0\rangle$:

$$
\begin{aligned}
p(|0\rangle) &= \langle\phi||0\rangle\langle 0||\phi\rangle \\
&= \frac{1}{2}\langle 0|\big(\langle\psi| + \langle\psi|U^{\dagger}\big) + \frac{1}{2}|1\rangle\big(\langle\psi| - U^{\dagger}\langle\psi|\big)(|0\rangle\langle 0|)\frac{1}{2}|0\rangle(|\psi\rangle + U|\psi\rangle) \\
&+\frac{1}{2}|1\rangle(|\psi\rangle - U|\psi\rangle) \\
&= \frac{1}{4}\langle 0|\big(\langle\psi| + \langle\psi|U^{\dagger}\big)\frac{1}{4}|0\rangle(|\psi\rangle + U|\psi\rangle) \\
&= \frac{1}{4}\big(\langle\psi| + \langle\psi|U^{\dagger}\big)(|\psi\rangle + U|\psi\rangle) \\
&= \frac{1}{4}\big(\langle\psi||\psi\rangle + \langle\psi|U^{\dagger}U|\psi\rangle + \langle\psi|U|\psi\rangle + \langle\psi|U^{\dagger}|\psi\rangle\big) \\
&= \frac{1}{4}(2 + 2\mathrm{Re}\langle\psi|U|\psi\rangle)
\end{aligned}
$$

Using the same process, for the second circuit one arrives at the final result $p(|0\rangle) = \frac{1}{2} + \frac{1}{2}\mathrm{Im}(\langle\psi|U|\psi\rangle)$.

Like the Pauli diagonalization method, the sampling complexity of this method scales as $O(1/\epsilon^2)$.

We now provide an example code performing the Hadamard test on a unitary operator. The operator in this example is the S gate, with the matrix representation

$$
S = \begin{pmatrix} 1 & 0 \\ 0 & i \end{pmatrix}
$$

This operator has been chosen for this example since the expectation values are purely real and imaginary for the quantum states $|0\rangle$ and $|1\rangle$, i.e.,

$$
\langle 0|S|0\rangle = 1
$$

$$
\langle 1|S|1\rangle = i
$$

In our example, we will rotate the state $|0\rangle$ toward the state $|1\rangle$ and back to $|0\rangle$ in increments using a rotation about Y. This can be achieved using the $R_Y(\theta)$ gate which has a matrix representation

$$
R_Y(\theta) = \begin{pmatrix} \cos\theta & -\sin\theta \\ \sin\theta & \cos\theta \end{pmatrix}
$$

In summary, we will compute the expectation value

$$
\langle\psi_\theta|S|\psi_\theta\rangle = \langle 0|R_Y(\theta)^{\dagger}SR_Y(\theta)|0\rangle
$$

using the Hadamard test and compare it with values computed classically. The results are plotted in Fig. 27.2:

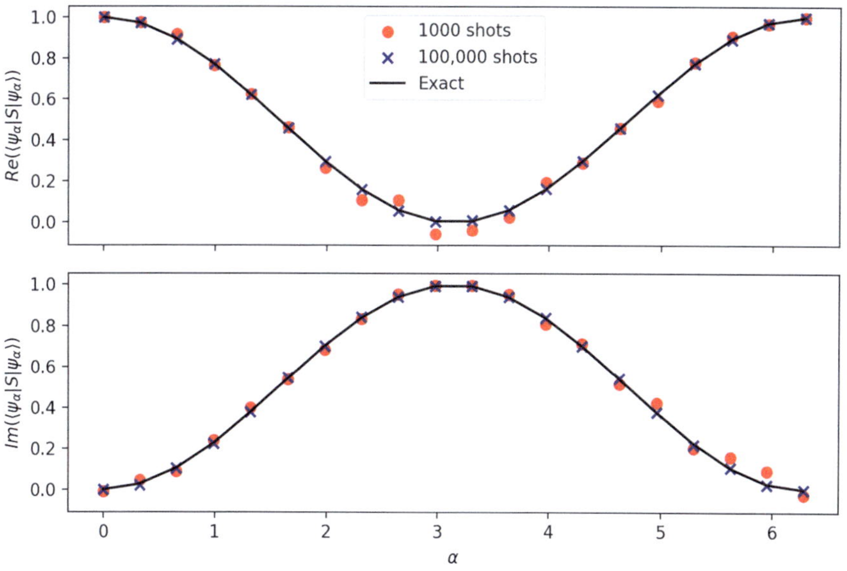

Fig. 27.2 Results of Hadamard test example code

```
#!/usr/bin/python3

import numpy as np
import matplotlib.pyplot as plt
from qiskit import QuantumCircuit, QuantumRegister, ClassicalReg-
ister
from qiskit.primitives import StatevectorSampler
from qiskit.circuit.library import SGate
from qiskit.circuit import Parameter

# Since we will be executing this circuit for various alpha values
# define alpha as a parameter
alpha = Parameter('alpha')
sweeps = 20
_alpha = np.linspace(0,np.pi*2, sweeps)
params = np.vstack([_alpha]).T

# Number of shots to estimate values
shots1 = 1000
shots2 = 100_000

# Set up classical and quantum registers
qregister = QuantumRegister(2)
```

```
cregister = ClassicalRegister(1,'classical')

# Create Hadamard test circuit for real part
re_circuit = QuantumCircuit(qregister,cregister)
re_circuit.ry(alpha,1) # Prepare |psi_alpha> state
re_circuit.h(0)
re_circuit.append(SGate().control(),[0,1])
re_circuit.h(0)
re_circuit.measure(0,0)
re_circuit.draw()

# Create Hadamard test circuit for imaginary part
im_circuit = QuantumCircuit(qregister,cregister)
im_circuit.ry(alpha,1) # Prepare |psi_alpha> state
im_circuit.h(0)
im_circuit.append(SGate().inverse(),[0])
im_circuit.append(SGate().control(),[0,1])
im_circuit.h(0)
im_circuit.measure(0,0)
im_circuit.draw()

# Sample circuits over various alpha values
sampler = StatevectorSampler()
# Define primitive unified blocks for real and imaginary circuits
pub1 = (re_circuit, params)
pub2 = (im_circuit, params)
# Run two jobs with different numbers of shots
job1 = sampler.run([pub1, pub2],shots=shots1)
job2 = sampler.run([pub1, pub2],shots=shots2)
# Extract p(0) from results for real and imaginary circuits
re1 = [(job1.result()[0].data.classical.get_counts(i)['0']/
shots1 * 2 - 1) for i in range(sweeps)]
im1 = [(job1.result()[1].data.classical.get_counts(i)['0']/
shots1 * 2 - 1) for i in range(sweeps)]
re2 = [(job2.result()[0].data.classical.get_counts(i)['0']/
shots2 * 2 - 1) for i in range(sweeps)]
im2 = [(job2.result()[1].data.classical.get_counts(i)['0']/
shots2 * 2 - 1) for i in range(sweeps)]

# Get the exact solution we expect to verify results
# Function returning Ry matrix for a theta value
def ry_matrix(theta):
        return np.array([[np.cos(theta/2), -np.sin(theta/
2)],[np.sin(theta/2), np.cos(theta/2)]])
```

```
# Create statevectors for various theta values
psi_theta = []
for theta in _alpha:
  psi_theta.append(ry_matrix(theta)@[[1],[0]])
# Compute expectation values using matrix-vector multiplication
s_matrix = [[1,0],[0,1j]]
evs = []
for psi in psi_theta:
  evs.append(psi.T.conj() @ s_matrix @ psi)

# Plot results
fig, (ax1, ax2) = plt.subplots(2, 1, sharex=True, sharey=True)
plt.xlabel(r'$\alpha$')
ax1.set_ylabel(r'$Re( \langle \psi_\alpha | S | \psi_\alpha \ran-
gle)$')
ax2.set_ylabel(r'$Im( \langle \psi_\alpha | S | \psi_\alpha \ran-
gle)$')
ax1.scatter(_alpha,re1,color='r',marker='o')
ax2.scatter(_alpha,im1,color='r',marker='o')
ax1.scatter(_alpha,re2,color='b',marker='x')
ax2.scatter(_alpha,im2,color='b',marker='x')
ax1.plot(_alpha,np.array(np.real(evs)).flatten(),color='k')
ax2.plot(_alpha,np.array(np.imag(evs)).flatten(),color='k')
ax1.legend(['1000 shots','100,000 shots','Exact'],loc=9)
plt.show()
```

Quantum Amplitude Estimation

Unlike the previous methods, the sample complexity of the amplitude estimation method scales as $O(1/\epsilon)$, achieving the so-called *Heisenberg limit*—the optimal scaling permitted in general by quantum mechanics.

To motivate the method, let's first revisit the Hadamard test from the previous section. Denoting a unitary that prepares $|\psi\rangle$ as U_ψ, and the circuit in (without the measurement operation) as U_{Had}, we combine the two to define

$$\mathcal{A} = U_{Had} U_\psi$$

Applying this operation to the state $|0\rangle\langle 0|^{\otimes n}$ we get

$$\mathcal{A}|0\rangle\langle 0|^{\otimes n} = |\phi\rangle = \frac{1}{2}|0\rangle(|\psi\rangle + U|\psi\rangle) + \frac{1}{2}|1\rangle(|\psi\rangle - U|\psi\rangle)$$

This state has a clear separation of "good" and "bad" states and can be restated as a sum

$$|\phi\rangle = \left|\phi_{good}\right\rangle + |\phi_{bad}\rangle$$

such that

$$\left\langle\psi_{good}|\psi\right\rangle = \left\langle\psi_{good}|\psi_{bad}\right\rangle = \sqrt{p}$$

$$\langle\psi_{bad}|\psi\rangle = \langle\psi_{bad}|\psi_{bad}\rangle = \sqrt{1-p}$$

where $\left|\phi_{good}\right\rangle = |0\rangle|\cdot\rangle$ and $|\phi_{bad}\rangle = |1\rangle|.\rangle$

Using the amplitude estimation procedure outlined in Chap. 24, Amplitude Amplification and Estimation., we may estimate p (and by extension $\frac{1}{2} + \frac{1}{2}\text{Re}(\langle\psi|U|\psi\rangle))$ up to precision ε with a complexity of $O(1/\epsilon)$.

The amplitude estimation procedure requires the construction of the reflection operator R_{good}. Recall that the amplitude amplification procedure requires marking the good states and the $|0\rangle|0\rangle^{\otimes n}$ state with a $-$ve sign, i.e., the operators S_{good} and S_0, respectively, need to be constructed.

For this problem we see a clear separation between the good and bad states from the ancilla qubit. This boils down to marking $|0\rangle|\cdot\rangle$ with a $-$ve sign. Therefore, we may construct the operator $\mathcal{S}_{good}$ as

$$S_{|\phi\rangle}|\phi\rangle = \big((XZX)\otimes(I)^{\otimes n}\big)|\phi\rangle = -\frac{1}{2}|0\rangle(|\psi\rangle + U|\psi\rangle) + \frac{1}{2}|1\rangle(|\psi\rangle - U|\psi\rangle)$$

The construction of S_0 is provided in Chap. 24: Amplitude Amplification and Estimation. Using the standard approach of phase estimation, we can now estimate $\frac{1}{2} + \frac{1}{2}\text{Re}(\langle\psi|U|\psi\rangle))$ up to precision ϵ. As shown in the previous section on the Hadamard test, we can extend this method to estimate $\frac{1}{2} + \frac{1}{2}\text{Im}(\langle\psi|U|\psi\rangle)$, completely determining the expectation value. This procedure has an overall sampling complexity of $O(1)$ and circuit depth $O(1/\epsilon)$. This leads to an overall complexity of $O(1/\epsilon)$, which is optimal and is known as the Heisenberg limit.

SWAP Test

In this chapter, we have covered various techniques for computing expectation values of the form $\langle\psi|H|\psi\rangle$. In some cases, however, we may be interested in computing the overlap between two quantum states:

$$\langle\psi|\phi\rangle$$

If the state preparation routines U_ψ and U_ϕ are known, one can compute the overlap using the Hadamard test by estimating the quantity:

$$\langle 0|U_\psi^\dagger U_{phi}|0\rangle$$

Another convenient way to estimate this overlap is to use a slightly modified version of the Hadamard test, known as either a Hadamard overlap test or a SWAP test. To build an intuition for this operation, consider two quantum registers, each containing n qubits.

$$|\psi\rangle|\phi\rangle$$

Applying the SWAP gate (or more precisely, a sequence of pairwise SWAPs) exchanges the content of the two registers:

$$U_{SWAP}|\psi\rangle|\phi\rangle = |\phi\rangle|\psi\rangle$$

We now apply the Hadamard test to this unitary operator U_{SWAP}, treating it as the controlled operation. This measures the expectation value:

$$\langle\psi|\langle\phi|U_{SWAP}|\psi\rangle|\phi\rangle = \langle\psi|\langle\phi||\phi\rangle|\psi\rangle = |\langle\psi|\phi\rangle|^2$$

Note that this yields the modulo squared value of the desired expectation value (and does not have an imaginary part), unlike the Hadamard test.

The circuit for the SWAP test is given in Fig. 27.3.

Note that the SWAP test may be extended to measure expectation values of the form

$$\langle\psi|U|\phi\rangle$$

Fig. 27.3 Quantum circuit for the SWAP test

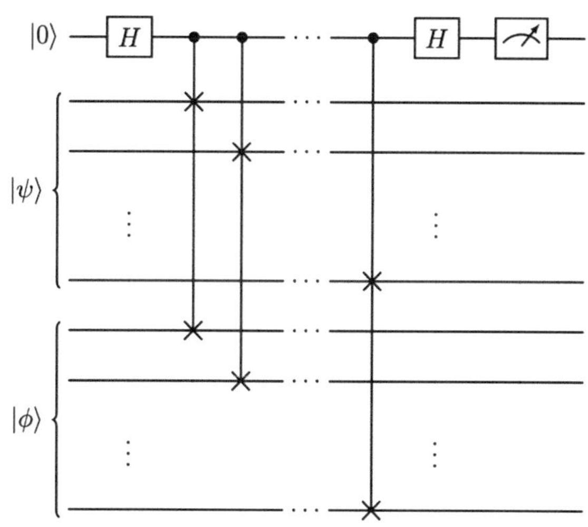

This can be achieved by performing a SWAP test of the form

$$\langle\psi|\langle\phi|(U \otimes U)U_{SWAP}|\psi\rangle|\phi\rangle$$

$$\langle\psi|\langle\phi|(U \otimes U)|\phi\rangle|\psi\rangle$$

$$|\langle\psi|U|\phi\rangle|^2$$

References

1. B. Reggio, N. Butt, A. Lytle, P. Draper, Fast partitioning of pauli strings into commuting families for optimal expectation value measurements of dense operators. Phys. Rev. A **110**(2), 022606 (2024). https://doi.org/10.1103/PhysRevA.110.022606
2. M. Kohda, R. Imai, K. Kanno, K. Mitarai, W. Mizukami, Y.O. Nakagawa, Quantum expectation-value estimation by computational basis sampling. Phys. Rev. Res. **4**(3), 033173 (2022). https://doi.org/10.1103/PhysRevResearch.4.033173

28 Hamiltonian Simulation Techniques

Hamiltonian simulation is the task of approximating the unitary operator $U \approx e^{-iHt}$ acting on a quantum state, where H is a Hermitian matrix and thus U is unitary. This subroutine is fundamental in quantum computing, with applications in quantum chemistry, condensed matter physics, and as a core routine in algorithms such as the HHL quantum linear system algorithm [1]. For quantum chemistry problems, the Jordan–Wigner [2] or Bravyi–Kitaev [3] transformations can be used to map the second quantized operators of an atom or molecule (a Fermionic Hamiltonian) to qubit operators and unitary operations of quantum computers [4].

Hamiltonian simulation encodes the time evolution of a quantum state governed by the Schrödinger equation:

$$\frac{d}{dt}|\Psi(t)\rangle = -iH(t)|\Psi(t)\rangle$$

where the Planck constant is absorbed into the Hamiltonian $H(t)$. For time-independent H, the solution is

$$|\Psi(t)\rangle = e^{-iHt}|\Psi(0)\rangle.$$

Thus, simulating a closed, time-independent quantum system is equivalent to solving a homogeneous first-order system of ordinary differential equations.

Several methods exist for Hamiltonian simulation, each suited to different settings and resource constraints. The most important approaches are product formulas, Taylor series expansions using linear combinations of unitaries, and quantum signal processing. These techniques offer a range of trade-offs in terms of efficiency, implementation overhead, and error scaling.

O. M. Raisuddin and S. De, *Quantum Computing for Engineers*,
https://doi.org/10.1007/978-3-032-03325-3_28

The Hamiltonian simulation problem for closed quantum systems is thus formally solved by the Schrodinger equation, which is a homogeneous system of first-order differential equations. The first quantum algorithm for this problem was proposed by [5] using the Trotter method. Subsequent work improved the query complexity to $O\left(d^2\|H_{\max}\|t\log\left(d^2\|H_{\max}\|/\epsilon\right)\right)$ by using a linear combination of unitaries arising from a truncated Taylor series [6]. The current state-of-the-art is quantum signal processing [7] which achieves a query complexity of $O\left(tdH_{\max}+\log\left(\frac{1}{\varepsilon}\right)/\log\log\left(\frac{1}{\epsilon}\right)\right)$ for d-sparse oracles, or $O\left(t\|H\|++\log\left(\frac{1}{\epsilon}\right)/\log\log\left(\frac{1}{\epsilon}\right)\right)$ with block-encoded oracles. Additional approaches, such as randomized evolution methods [8, 9], have also been developed.

Trotter Methods

The Hamiltonian simulation problem is relatively straightforward to solve using Trotter methods. This requires decomposing the Hamiltonian into a sum of easily exponentiated summands H_j:

$$H = \sum_j H_j$$

These summands H_j are often chosen to be Pauli strings, since various physical Hamiltonians can be expressed in this form. This decomposition can be used in various Trotter formulas to obtain an approximation of e^{-iHt}. The order of the Trotter formula impacts the simulation errors. Grouping commuting Pauli strings often leads to lower error rates. Trotter methods are discussed in detail in Chap. 21: Trotterization.

Taylor Series Approximation

The Taylor series approach approximates a matrix exponential with a matrix polynomial, which can be implemented as a sum of monomials using the linear combination of unitaries (LCU) technique. Since $U = e^{-iHt}$ is analytic, it can be approximated by a truncated Taylor series. This method starts by decomposing H as a sum of unitaries H_l [10], similar to the product formula approach:

$$U = e^{-iHt} = \left(e^{-iHt/r}\right)^r = (U_r)^r$$

where

$$U_r \approx \hat{U}_r = \sum_{k=0}^{K} \left(\sum_{l_1,\ldots,l_k=1}^{L} \frac{(-it/r)^k}{k!} \alpha_{l_1} \ldots \alpha_{l_k} H_{l_1} \ldots H_{l_k} \right).$$

Choosing $K = O\left(\frac{\log(\tau/\epsilon)}{\log\log(\tau/\epsilon)}\right)$ achieves precision $\left\|U - (\hat{U}_r)^r\right\|_2 \leq \epsilon$, yielding exponential improvement over Trotter methods. If H_l are Pauli strings, their products $H_{l_1} \ldots H_{l_k}$ remain unitary. Therefore, the double summation may be implemented using the LCU subroutine, which requires ancilla qubits and has a non-zero probability of failure. We note that if the LCU subroutine is implementing (approximately) unitary operations its success probability can be boosted to $O(1)$ using a subroutine known as Oblivious Amplitude Amplification [11], which—unlike regular amplitude amplification—boosts amplitudes linearly rather than quadratically.

The overall algorithm simulates a d-sparse Hamiltonian H for time t to a precision of ε with $O\left(\tau \frac{\log^2(\tau/\epsilon)}{\log\log(\tau/\epsilon)}\right)$ queries to H, where $\tau = d^2\|H_{\max}\|t$, which is near-optimal in time. However, this approach requires additional ancilla qubits and a large number of controlled operations. The quantum signal processing method, presented in the next section, requires significantly fewer ancillae and controlled operations. Neither method is practical on pre-fault-tolerant quantum computers.

Quantum Signal Processing

The quantum signal processing (QSP) method also approximates a matrix exponential with a matrix polynomial, but instead of summing monomial terms, it uses the QSP framework, requiring only two additional ancilla qubits. QSP achieves optimal scaling by simulating a d-sparse Hamiltonian for time t with error ϵ using $O\left(td\|H_{\max}\| + \frac{\log(1/\epsilon)}{\log\log(1/\epsilon)}\right)$ queries to H [7].

The method employs a complex polynomial approximation of e^{-iHt}. By Euler's formula,

$$e^{-iHt} = \cos(iHt) - \sin(iHt)$$

$\cos(iHt)$ and $\sin(iHt)$ are ϵ-approximated using the Jacobi–Anger expansion:

$$\cos(xt) \approx P_{\cos}(x) = J_0(t) + 2\sum_{k=1}^{k'}(-1)^k J_{2k}(t)T_{2k}(x)$$

$$\sin(xt) \approx P_{\sin}(x) = 2\sum_{k=0}^{k'}(-1)^k J_{2k+1}(t)T_{2k+1}(x)$$

where $J_i(x)$ is the ith order Bessel function and $T_i(x)$ is the ith Chebyshev polynomial of the first kind with a choice of $k' = \frac{1}{2}r\left(\frac{e}{2}|t|, \frac{5}{4}\epsilon\right)$. Implementing these polynomials using QSP requires choosing a truncation error of ε/4 and rescaling by a factor of $\frac{1}{1+\epsilon/4}$ to ensure that both $\|P_{\cos}(x) + P_{\sin}(x)\| \leq 1$ and $\|P_{\cos}(x) + P_{\sin}(x)\| - e^{ix} \leq \epsilon$. This approach achieves optimal scaling in t by simulating a d-sparse Hamiltonian for time t with error ε using $O\left(td\|H_{\max}\| + \frac{\log(1/\epsilon)}{\log\log(1/\epsilon)}\right)$ queries to H [7].

Since the QSP implementation of $\cos(iHt)$ and $\sin(iHt)$ is valid for $t \geq 0$ and positive-definite H, more general cases can be handled by using block-encoding of H and defining $H_+ = \frac{1}{2}(H/\alpha + I)$, where α is the subnormalization factor of the block encoding. The operator H_+ is positive-definite and the evolution of $e^{-2iH_+\alpha t}$ is equivalent to e^{-iHt} up to a global phase factor.

The techniques outlined in this chapter are categorized as digital Hamiltonian simulation, since they realize the action of the Hamiltonian as a sequence of discrete steps. Analog methods directly map the Hamiltonian to an equivalent physical system. These methods are not generally applicable to gate-based quantum computers and are beyond the scope of this book.

We discuss the solution of non-unitary systems of ordinary differential equations in more detail in Chap. 32, Quantum Ordinary Differential Equation Algorithms: Block-Matrix Algorithms, and Chap. 33, Quantum Ordinary Differential Equation Algorithms: Time-Marching Algorithms.

The no-fast-forwarding theorem places a lower bound on the asymptotic complexity of Hamiltonian simulation, establishing that Hamiltonian simulation cannot be performed in sublinear time in general. The proxy problem for this proof is the problem of computing the parity of a string of bits [12].

Hamiltonian simulation has great potential for speeding up scientific and engineering computations. Homogeneous linear systems of differential equations can be transformed into a system of first-order differential equations using a transformation known as "Schrodingerization," which can then be solved using Hamiltonian simulation techniques. The speedups for quantum chemistry itself can be immense, allowing ab initio computation of larger systems with higher accuracy. While density functional theory (DFT) is a widely used semi-empirical method for ab initio calculations, it relies on controlled approximations and usually can only provide ground-state solutions, limiting its application. Direct quantum simulations can provide more accurate results and can be used for multiscale modeling to unlock the interesting physics arising from the excited states of molecules and condensed matter, such as crystals.

To realize the exponential speedups offered by quantum computers for Hamiltonian simulation, efficient implementations of oracles or block-encodings need to be developed for problems of practical interest. Furthermore, error rates need to be low and coherence times must be large in quantum devices to extract useful results.

References

1. A.W. Harrow, A. Hassidim, S. Lloyd, Quantum algorithm for linear systems of equations. Phys. Rev. Lett. **103**(15), 150502 (2009). https://doi.org/10.1103/PhysRevLett.103.150502
2. P. Jordan, E. Wigner, Über das Paulische äquivalenzverbot. Z. Physik **47**(9–10), 631–651 (1928). https://doi.org/10.1007/BF01331938
3. S.B. Bravyi, A.Y. Kitaev, Fermionic quantum computation. Ann. Phys. **298**(1), 210–226 (2002). https://doi.org/10.1006/aphy.2002.6254

4. J.D. Whitfield, J. Biamonte, A. Aspuru-Guzik, Simulation of electronic structure Hamiltonians using quantum computers. Mol. Phys. **109**(5), 735–750 (2011). https://doi.org/10.1080/00268976.2011.552441
5. C. Coppersmith, An approximate Fourier transform useful in quantum factoring (IBM Research Division, RC, 19642,1994). https://doi.org/10.48550/arXiv.quant-ph/0201067
6. A.I. Google Quantum et al., Exponential suppression of bit or phase errors with cyclic error correction. Nature **595**(7867), 383–387 (2021). https://doi.org/10.1038/s41586-021-03588-y
7. G.H. Low, I.L. Chuang, Optimal Hamiltonian simulation by quantum signal processing. Phys. Rev. Lett. **118**(1), 010501 (2017). https://doi.org/10.1103/PhysRevLett.118.010501
8. C.-F. Chen, H.-Y. Huang, R. Kueng, J.A. Tropp, Concentration for random product formulas. PRX Quantum **2**(4), 040305 (2021). https://doi.org/10.1103/PRXQuantum.2.040305
9. A.M. Childs, A. Ostrander, Y. Su, Faster quantum simulation by randomization. Quantum **3**, 182 (2019). https://doi.org/10.22331/q-2019-09-02-182
10. D.W. Berry, A.M. Childs, R. Cleve, R. Kothari, R.D. Somma, Simulating Hamiltonian dynamics with a truncated Taylor series. Phys. Rev. Lett. **114**(9), 090502 (2015). https://doi.org/10.1103/PhysRevLett.114.090502
11. D.W. Berry, A.M. Childs, R. Cleve, R. Kothari, R.D. Somma, Exponential improvement in precision for simulating sparse Hamiltonians, in *Proceedings of the Forty-Sixth Annual ACM Symposium on Theory of Computing, in STOC '14* (Association for Computing Machinery, New York, NY, USA, 2014), pp. 283–292. https://doi.org/10.1145/2591796.2591854
12. D.W. Berry, G. Ahokas, R. Cleve, B.C. Sanders, Efficient quantum algorithms for simulating sparse Hamiltonians. Commun. Math. Phys. **270**(2), 359–371 (2007). https://doi.org/10.1007/s00220-006-0150-x

Eigenvalue Problems

29

The estimation of eigenvalues is a significant application of quantum computers, with implications for quantum chemistry (ground and excited states and their energies), materials science, and various problems in physics. Among some of the well-known algorithms are the Variational Quantum Eigensolver (VQE), quantum Krylov subspace methods, and Quantum Phase Estimation (QPE). VQE and related variational algorithms are discussed in Chap. 36: Notable Variational Algorithms: VQE, QAOA, and VQLS. QPE has been presented in Chap. 20: Quantum Phase Estimation. This chapter focuses on quantum Krylov methods for eigenvalue problems.

Krylov Methods

Consider a problem where a quantum state $|\psi_0\rangle$ can be time-evolved by the Hamiltonian H as e^{-iHt}. By choosing $t = \{t_0, t_1, ..t_m\}$ where $t_{i+1} = t_i + \Delta t$, one can build a Krylov subspace

$$\left\{\langle\psi_0|, e^{-iH\Delta t}|\psi_0\rangle, e^{-iH2\Delta t}|\psi_0\rangle, e^{-iH3\Delta t}|\psi_0\rangle, \ldots\right\} = \{\langle|\psi_0\rangle, |\psi_1\rangle, |\psi_2\rangle, |\psi_3\rangle, \ldots\}$$

Expectation values of the form $\langle\psi_i|H\left|\psi_j\right\rangle$ and $\left\langle\psi_i|\psi_j\right\rangle$ are then estimated to form the generalized eigenvalue problem:

$$\tilde{H}\phi_i = \lambda_i S\phi_i$$

O. M. Raisuddin and S. De, *Quantum Computing for Engineers*,
https://doi.org/10.1007/978-3-032-03325-3_29

where

$$S = \begin{pmatrix} \langle \psi_i | \psi_i \rangle & \langle \psi_i | \psi_j \rangle & \cdots \\ \langle \psi_j | \psi_i \rangle & \langle \psi_j | \psi_j \rangle & \\ \vdots & & \ddots \end{pmatrix}$$

and

$$\tilde{H} = \begin{pmatrix} \langle \psi_0 | H | \psi_0 \rangle & \langle \psi_0 | H | \psi_1 \rangle & \cdots \\ \langle \psi_1 | H | \psi_0 \rangle & \langle \psi_1 | H | \psi_1 \rangle & \\ \vdots & & \ddots \end{pmatrix}$$

Note that S and $\tilde{H}$ are symmetric. The elements of S and $\tilde{H}$ may be computed using techniques presented in Chap. 27: Expectation Value Estimation.

The lowest generalized eigenvalue corresponds to the minimum eigenvalue in the subspace spanned by $\{|\psi_i\rangle\}$. However, the generalized eigenvalue problem arising from this subspace construction method is often ill-conditioned, and a dynamical mode decomposition (DMD) method has been proposed to improve numerical stability [1, 2].

In some cases, a good approximation of the relevant subspace may already be available, and the subspace vectors $|\psi_1\rangle, \ldots$ can be chosen directly rather than generated via the Krylov method. More generally, an arbitrary subspace can be constructed by other means as well.

Approximate eigenvalue estimation and eigenstate preparation are central applications of quantum computers in physics and chemistry and constitute an active area of research. Other notable techniques, such as QCELS [3], quantum imaginary time evolution [4], and sample-based quantum diagonalization [5], are beyond the scope of this book.

References

1. Y. Shen et al., Estimating eigenenergies from quantum dynamics: a unified noise-resilient measurement-driven approach (2023). arXiv:2306.01858. https://doi.org/10.48550/arXiv.2306.01858
2. Y. Shen et al., Efficient measurement-driven eigenenergy estimation with classical shadows (2024). arXiv:2409.13691. https://doi.org/10.48550/arXiv.2409.13691
3. Z. Ding, L. Lin, Even shorter quantum circuit for phase estimation on early fault-tolerant quantum computers with applications to ground-state energy estimation. PRX Quantum **4**(2), 020331 (2023). https://doi.org/10.1103/PRXQuantum.4.020331
4. M. Motta et al., Determining eigenstates and thermal states on a quantum computer using quantum imaginary time evolution. Nat. Phys. **16**(2), 205–210 (2020). https://doi.org/10.1038/s41567-019-0704-4
5. J. Robledo-Moreno et al., Chemistry beyond exact solutions on a quantum-centric supercomputer (2024). https://doi.org/10.48550/ARXIV.2405.05068

Quantum Linear System Algorithms: Direct Methods

30

Quantum linear system algorithms (QLSA) are central to the potential utility of quantum computing for scientific and engineering applications, as linear systems are ubiquitous in these fields. While classical algorithms for solving an N-dimensional linear system $Ax = b$ require at least $O(N)$ time to produce the full solution vector x, many engineering and scientific applications only need a few specific properties, such as the peak stress in a structure or the lift coefficient of an airfoil. QLSAs promise to bypass this bottleneck by directly preparing a quantum state whose amplitudes encode the solution, often in time polylogarithmic in N, potentially enabling the efficient extraction of these observables.

In this chapter, we focus on direct QLSAs, outlining the foundational Harrow–Hassidim–Lloyd (HHL) algorithm [1] and its subsequent refinements—Linear Combination of Unitaries (LCU) [2], and Quantum Signal Processing (QSP/QSVT) [3], discussing their underlying assumptions, computational complexity, and practical limitations.

Formally, QLSAs solve the Quantum Linear System Problem (QLSP) [1] as follows.

Definition (*Quantum Linear System Problem*) Given a normalized matrix $A \in \mathbb{C}^{N\times N}$ with $\|A\| = 1$, a vector $b \in \mathbb{C}^N$, oracle access to the entries of A, and the ability to prepare a quantum state $|b\rangle = \frac{\sum_i b_i|i\rangle}{\|\sum_i b_i|i\rangle\|_2}$, the task is to prepare a quantum state $|\tilde{x}\rangle$, such that $\||\tilde{x}\rangle - |x\rangle\|_2 \le \epsilon$, where $|x\rangle = \frac{\sum_i x_i|i\rangle}{\|\sum_i x_i|i\rangle\|_2}$ and $x = A^{-1}b$.

O. M. Raisuddin and S. De, *Quantum Computing for Engineers*,
https://doi.org/10.1007/978-3-032-03325-3_30

Table 30.1 A summary of notable QLSAs

Algorithm	Complexity	Pros	Cons	Notes
HHL [1] (2007)	$O\big(d^2\kappa^2 \log(N)/\epsilon\big)$	1 ancilla Short circuit possible	High error	$\kappa^2 \to \kappa \log \kappa$ using VTAA [4]
LCU [2] (2017)	$O\big(\kappa^2 \text{poly}\log(\kappa N/\epsilon)\big)$	Many ancillae	Low error Complex circuit	$\kappa^2 \to \kappa \log \kappa$ using AA [2]
QSP/QVST [3] (2021)	$O(\kappa \log(\kappa N/\epsilon))$	Few ancillae, near-optimal scaling	QSP sequence Φ required	Φ can be reused
PD-QLSA [5] (2021)	$O\big(\sqrt{\kappa} \log(\kappa N/\epsilon)\big)$	Near-optimal $\sqrt{\kappa}$ scaling classical scaling	Only for SPD systems	Requires upper bound on $\|A\|_2$
DAT [6] (2022)	$O(\kappa \log(N/\epsilon))$	Optimal κ scaling	Requires construction of a Hamiltonian for evolution	

Notes:

- Most QLSA, including HHL, LCU, and quantum signal processing methods, require A to be Hermitian, or they rely on an efficient reduction of the general case to an equivalent Hermitian problem (e.g., using the Hermitian dilation of A). For non-Hermitian A, this embedding increases the problem size but enables the use of these algorithms.
- Due to the normalization constraint of quantum states, $|x\rangle$ is proportional to the classical solution x, and its amplitudes encode the solution information. Unlike classical approaches, the full solution vector is not output; instead, quantum algorithms enable the efficient extraction of relevant observables or sampling.

Since quantum algorithms can operate on the entire state efficiently, QLSAs can yield exponential speedups in N for suitable QLSPs [1]. A summary of these methods is presented in Table 30.1. As discussed here, QLSAs are quantum analogs of direct classical solvers—they do not depend on an initial guess or leverage problem-specific structure. Quantum iterative methods, which offer complementary strategies, are addressed in the following chapter.

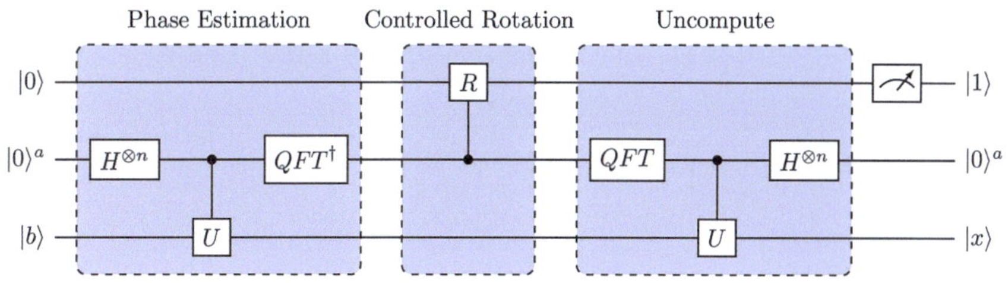

Fig. 30.1 A quantum circuit demonstrating the HHL algorithm

HHL Algorithm

The Harrow–Hassidim–Lloyd (HHL) algorithm was the first to demonstrate an exponential advantage in system size for linear systems of equations, with complexity $O\big(\kappa^2 d^2 \log(N)/\epsilon\big)$ where d is the maximum number of non-zero entries in any row or column of A, and κ is the condition number. HHL leverages Hamiltonian simulation to apply controlled e^{-iAt} operations on the input state $|b\rangle$. Since A and e^{-iAt} share the same eigenvectors, the eigenvalues of e^{-iAt} are kicked back to the control register via quantum phase estimation. Using the quantum Fourier transform, eigenvalues are encoded in the amplitudes of the control qubits, and a controlled rotation on an ancilla is used to invert the eigenvalues. The computation is then reversed to disentangle the registers, and the ancilla is measured. Measurement of the ancilla in the desired state indicates a successful solution.

Since the algorithm relies on phase estimation, it suffers from poor scaling in precision. Variable-time amplitude amplification can improve the scaling with respect to the condition number κ to linear [4].

An overview schematic of HHL circuit is given in Fig. 30.1.

Given subsequent improvements beyond HHL, we do not discuss them in detail here; instead, we proceed directly to more advanced algorithms.

LCU-Based Methods

Similar to Taylor series methods discussed for Hamiltonian simulation, one can approximate the matrix function $f(x) = \frac{1}{x}$ (since we seek an approximation of $f(A) = A^{-1}$) using polynomial or Fourier expansions. However, unlike Hamiltonian simulation, which directly approximates a matrix exponential e^A, matrix inversion requires some additional considerations.

First, the matrix must be normalized such that $\|A\| \leq 1$. This ensures that the singular values of A lie in the disjoint interval $[-1, -1/\kappa] \cup [1/\kappa, 1]$. This can be achieved by normalizing the system matrix using an upper bound $\alpha \geq \|A\|$ (note that $\alpha \neq \|A\|$ will require an interval $[-1, -\|A\|/\alpha\kappa] \cup [\|A\|/\alpha\kappa, 1]$).

Second, since $\lim_{x\to 0} f(x) \to \infty$, $f(x)$ must be approximated over a disjoint interval $\left[-1, -\frac{1}{\kappa}\right] \cup \left[\frac{1}{\kappa}, 1\right]$. Thus, a standard Taylor series approximation or a Chebyshev approximation is not sufficient.

Finally, for $f(A) = A^{-1}$ to hold, A must be Hermitian. In case A is not Hermitian, one may use Hermitian dilation to instead solve the equivalent linear system:

$$\begin{pmatrix} 0 & A \\ A^\dagger & 0 \end{pmatrix} \begin{pmatrix} 0 \\ x \end{pmatrix} = \begin{pmatrix} b \\ 0 \end{pmatrix}$$

The LCU algorithm [2] instead uses either a Chebyshev polynomial approximation $P(x) = \Sigma_k T_k(x) \approx 1/x$, where T_k are Chebyshev polynomials of the first kind, or a Fourier approximation $G(x) = \Sigma_j e^{it_j x} \approx 1/x$, where $t_j \in \mathbb{R}$, over the interval $x \in I_P : [-1, -1/\kappa] \cup [1/\kappa, 1]$, where κ is the condition number of the system. A filter function is used to handle the disjoint interval, details of which can be found in [2].

The matrix polynomial approximation $P(A)$ or Fourier approximation $G(A)$ is applied to the state $|b\rangle$ to obtain the approximate solution $A^{-1}b \approx P(A)b$ (or $A^{-1}b \approx G(A)b$). Given a desired precision $\left\|P(x) - \frac{1}{x}\right\|_{\max x \in I_P} \leq \epsilon$ or $\left\|G(x) - \frac{1}{x}\right\| \leq \epsilon$, $P(x)$ is a polynomial of degree $O(\kappa \log(\kappa/\epsilon))$ and $G(x)$ is a Fourier expansion with $O\left(\kappa\sqrt{\log(\kappa/\epsilon)}\right)$ terms. The terms $T_k(x)$ can be formed either using matrix multiplications through qubitization or quantum walks (an equivalent approach for forming Chebyshev polynomials on gate-based quantum computers), and $G(x)$ can be formed using Hamiltonian simulation.

However, the application of a linear combination of unitaries incurs additional overhead in the form of ancilla qubits. The overall algorithm requires $O\left(d\kappa \operatorname{poly}\log\left(\frac{d\kappa}{\epsilon}\right)\right)$ queries to a sparse access oracle for A and requires $O\left(d\kappa \operatorname{poly}\log\left(\frac{d\kappa N}{\epsilon}\right)\right)$ resources. Although its implementation is rather involved, the LCU QLSA is a seminal development for QLSAs with exponentially improved scaling in ϵ.

Quantum Signal Processing

Similar to its application in Hamiltonian simulation, the QSP method [3] addresses the need for matrix function approximation without requiring ancilla qubits. QSP works by finding a sequence of phase angles Φ, corresponding to the desired polynomial and a block-encoding of the matrix, given bounds on the condition number κ and the desired precision ϵ. The sequence Φ is not specific to the problem and can be reused for any other problem as long as $\kappa_{\text{new}} \leq \kappa$ and $\epsilon_{\text{new}} \leq \epsilon$. Given a block-encoded oracle for A, the query complexity of the algorithm is $O\left(\kappa \log\left(\frac{\kappa}{\epsilon}\right)\right)$, and it requires $O\left(\kappa \log\left(\frac{\kappa N}{\epsilon}\right)\right)$ resources.

For symmetric positive-definite systems, it is possible to achieve the $\sqrt{\kappa}$ scaling of classical solvers [5]. This is done by first defining the alternative polynomial approximation $Q(y) \approx y = \frac{1}{1-x}$ over the interval $y \in I_Q : [-1, 1/\kappa]$, and defining $B = I - \eta A$ where η is chosen s.t. $\|B\| \leq 1$. Given a desired precision $\|Q(x) - \frac{1}{x}\|_{\max x \in I_Q} \leq \epsilon$, $Q(x)$ is a polynomial of degree $O\big(\sqrt{\kappa}\log(\kappa/\epsilon)\big)$. The matrix polynomial approximation $Q(B)$ is then applied to the quantum state $|b\rangle$.

Below is an example of a QLSA implementation using quantum signal processing in Qiskit. The code inverts an anti-diagonal matrix A of the form:

$$A = \begin{pmatrix} & & & 1 \\ & & 0.95 & \\ & \cdot^{\cdot^{\cdot}} & & \\ -0.95 & & & \\ -1 & & & \end{pmatrix}$$

whose inverse is

$$A^{-1} = \begin{pmatrix} & & & -1 \\ & & -1/0.95 & \\ & \cdot^{\cdot^{\cdot}} & & \\ 1/0.95 & & & \\ 1 & & & \end{pmatrix}.$$

Figure 30.2 shows the agreement of the QSP polynomial with $1/x$ with $\epsilon \leq 10^{-6}$. The phase factors Φ are obtained from the QSPPACK library:

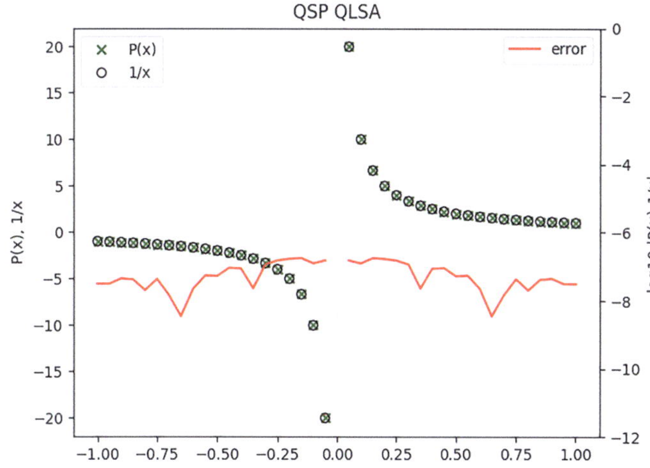

Fig. 30.2 Output of the example code implementing the QSP QLSA. The code demonstrates the agreement of the singular values of $P(A) \approx A^{-1}$ with x^{-1}

```
#!/usr/bin/python3

import qiskit
from qiskit import QuantumCircuit, QuantumRegister
from qiskit import ClassicalRegister
from qiskit.quantum_info.operators import Operator
from qiskit_aer import Aer
import numpy as np
from scipy.linalg import fractional_matrix_power
from scipy.io import loadmat
from copy import deepcopy
from matplotlib import pyplot as plt

# Define the linear system:
# As an example, solve a matrix with numbers +,- 1...0.05 on the anti-
diagonal,
# which is a non-Hermitian matrix
step = 0.05
A = np.diag(np.concatenate( (np.arange(-1,0,step),np.arange(step,
1+step,step))))
# Take anti-transpose of A to demonstrate the non-Hermitian matrix
case.
A = np.flipud(A)
# Save matrix to check againt classical solution later
A_orig = deepcopy(A)
# Normalize A. If an upper bound is known, use that instead.
# A = A/np.linalg.norm(A)

b = np.ones(np.shape(A)[1])
# Turn b into a quantum state
b = b/np.linalg.norm(b,2)
b_orig = deepcopy(b)

# Hermitian Dilation: only if A is not Hermitian
if np.any(A != A.conj().T):
   A = np.block([
      [np.zeros(np.shape(A)),A],
      [A.conj().T,np.zeros(np.shape(A))]
   ])
   b = np.block([
      b,
      np.zeros(np.shape(b))
   ])
   HD = True
```

```
else:
  HD = False

# The matrix A needs to padded to some 2^n to enable block-encoding
if np.size(A)>1:
  A_num_qubits = int(np.ceil(np.log2(np.shape(A)[0])))
  padding_size = 2**A_num_qubits - np.shape(A)[0]
  if padding_size > 0:
    A = np.block([
      [A, np.zeros([np.shape(A)[0],padding_size])],
                    [np.zeros([padding_size,np.shape(A)[0]]),
np.zeros([padding_size,padding_size])]
    ])
else:
  A_num_qubits = 1
  padding_size = 1
  A = np.array([[A,0],[0,0]])
# Similarly, pad b
b = np.pad(b,(0,padding_size))

# Define the block-encoding of the matrix A
# If you have an efficient circuit to realize U_A (or O_A), use it
here
U_A = np.block([
    [A    ,    -fractional_matrix_power(np.eye(np.shape(A)[0]) -
np.linalg.matrix_power(A,2),0.5)],
             [fractional_matrix_power(np.eye(np.shape(A)[0])    -
np.linalg.matrix_power(A,2),0.5), A]
])
# We also need to get the block-encoding size, i.e. m, used to encode
A in U_A
m = int(np.log2(np.shape(U_A)[0]) - A_num_qubits)
U_A_num_qubits = int(np.log2(np.shape(U_A)[0]))

# Create the operator U_A in Qiskit
operatorA = Operator(U_A)
# Create the three registers for QSP:
# 1) 1 Z rotation qubit
# 2) m block-encoding ancillae
# 3) register for b
register_1 = QuantumRegister(size = 1, name = '|0>')
register_2 = QuantumRegister(size = m, name = '|0^m>')
register_3  =  QuantumRegister(size  =  U_A_num_qubits-m,  name  =
'|\phi>')
```

```
# Create a rotation circuit in the block-encoding basis
def CR_phi_d(phi, d, register_1, register_2):
    circuit = QuantumCircuit(register_1,register_2,name = 'CR_(
\phi \tilde {})'.format(d))
   circuit.cx(register_2,register_1,ctrl_state=0)
   circuit.rz(phi*2, register_1)
   # Done this way for numerical stability
   circuit.z(register_1)
   circuit.cx(register_2,register_1,ctrl_state=0)
   return circuit

# Load QSP angles
# These angles can be obtained from the QSPPACK package
phi_angles   =   np.array(   loadmat('phi_kappa_80_pts_8000_deg_
1999.mat')).item()['phi_proc']

phi_tilde_angles = np.zeros(np.shape(phi_angles))
phase_angles = phi_angles.reshape(phi_angles.shape[0])

# Create QSP circuit
QSP_circuit = QuantumCircuit(register_1, register_2, register_3,
name = 'QSP')
# Initialize state |b>. If you have an efficient implementation for
b, it goes here
QSP_circuit.initialize(b,list(reversed(register_3)))

# First Hadamard the ancilla qubit since we want Re(P(A))
QSP_circuit.h(register_1)
# Note: QSPPACK produces symmetric phase angles, so reversing phase
angles is unnecessary
for d, phi in reversed(list(enumerate(phase_angles))):
     QSP_circuit  =  QSP_circuit.compose(CR_phi_d(phi,d,register_
1,register_2))
   if d>(0):
      # The endianness of the bits matters. Need to change the order
of the bits
      if d%2:
            QSP_circuit.append(operatorA.adjoint(),list(reversed
(register_3[:])) + register_2[:])
   else:
      QSP_circuit.append(operatorA,list(reversed(register_3[:]))
+ register_2[:])

# Apply the final Hadamard gate
```

```
QSP_circuit.h(register_1)
# Account for little vs. big endian
QSP_circuit = QSP_circuit.reverse_bits()

# Run statevector simulator
solver='statevector'
backend = Aer.get_backend('statevector_simulator',precision =
"double")
job = backend.run(QSP_circuit, shots=0)

# Extract relevant portion of statevector
QSP_statevector = job.result().get_statevector()
if HD:
            P_A_b    =    np.real(QSP_statevector.data[int(b_
orig.shape[0]):(2*b_orig.shape[0])])
else:
  P_A_b = np.real(QSP_statevector.data[0:b.shape[0]])
P_A_b = P_A_b/np.linalg.norm(P_A_b)

# Get expected result using classical solver
expected_P_A_b = np.linalg.solve(A_orig,b_orig)
expected_P_A_b = expected_P_A_b/np.linalg.norm(expected_P_A_b)

# Plot QSP polynomial
x = np.flipud(A_orig).diagonal()
fig, ax1 = plt.subplots()
ax1.set_title('QSP QLSA')
ax1.scatter(x,P_A_b/P_A_b[-1],marker='x',c='g')
ax1.scatter(x,expected_P_A_b/expected_P_A_b[-
1],marker='o',facecolors='none', edgecolors='k')
ax1.set_ylabel('P(x), 1/x')
plt.legend(['P(x)','1/x'],loc = 2)
ax2 = ax1.twinx()
ax2.plot(x[:x.size//2],np.log10(np.abs((P_A_b[:x.size//
2]-expected_P_A_b[:x.size//2])/expected_P_A_b[-1])),'r')
ax2.plot(x[x.size//2:],np.log10(np.abs((P_A_b[x.size//
2:]-expected_P_A_b[x.size//2:])/expected_P_A_b[-1])),'r')
ax2.set_ylim(bottom=-12, top=0)
ax2.set_ylabel('log10 |P(x)-1/x|')
plt.legend(['error'],loc = 1)
plt.xlabel('x')
plt.show()
```

We finally note that optimal scaling in κ has been achieved by Orsucci and Dunjko [5], with a query complexity of $O\left(\kappa \log\left(\frac{1}{\epsilon}\right)\right)$ using the adiabatic theorem.

The linear scaling in κ for general linear systems implies that QLSAs can only provide an exponential speedup for problems with $\kappa = O(\text{poly} \log N)$. Practical problems of interest rarely exhibit such scaling. To address this, [7] has proposed a quantum sparse approximate inverse (SPAI) preconditioner, which constructs the preconditioner by solving a least-squares problem on each row of the matrix. However, the SPAI preconditioner may be inefficient for many problems [8], and a complete implementation of the algorithm is unavailable. Alternatively, a circulant matrix preconditioner has been suggested in [9], but its complexity depends on the condition number of the preconditioner $\kappa(M)$ and the product $\kappa\left(M^{-1}A\right)$. In the worst case, $\kappa(M) \geq \kappa(A)$ and $\kappa(M), \kappa\left(M^{-1}A\right) \approx \sqrt{\kappa(A)}$. A Laplacian preconditioner has been proposed by Golden et al. [8] for hydrological subsurface flow to reduce the condition number to at most $O\left(\sqrt{N}\right)$. For certain QLSPs of the form $(A + B)|x\rangle = |b\rangle$ with $\|A\| \ll \|B\|$, Tong et al. [10] proposed a fast-inversion procedure, relevant for single-particle Green's functions in quantum many-body systems.

Several state-of-the-art quantum algorithms for inhomogeneous ordinary differential equations rely on QLSAs as an intermediate step, which enables their exponential speedups [11, 12]. Numerical solutions of partial differential equations can greatly benefit from QLSAs if the condition number can be controlled. An iterative approach has been proposed by Raisuddin and De [13] to manage the condition number of the systems of equations, with possible application to a quantum multigrid method or a quantum domain decomposition method. We discuss quantum algorithms for ODEs and PDEs in Chap. 32, Quantum Ordinary Differential Equation Algorithms: Block-Matrix Algorithms, Chap. 33, Quantum Ordinary Differential Equation Algorithms: Time-Marching Algorithms, and Chap. 34, Quantum Partial Differential Equation Algorithms.

Since the QLSA prepares a quantum state encoding the solution, it also raises the question of computing practically relevant properties of the output. Clader et al. [7] proposed using QLSAs to compute the electromagnetic scattering cross-section of an arbitrary target. Montanaro and Pallister [14] noted that while the QLSA provides exponential speedup in state preparation, extracting classical properties from the quantum state yields only polynomial speedup for finite element problems, with the speedup increasing for higher dimensional cases.

A further analysis in [15] compared classical and quantum methods for the heat equation, showing that direct QLSA application is never faster than classical algorithms. However, an approach based on amplification and random walks can yield quadratic speedup for $d \geq 2$, and even greater speedup for higher d. The heat equation is especially relevant since it discretizes the Laplacian operator, which appears in a broad class of PDEs [16].

References

1. A.W. Harrow, A. Hassidim, S. Lloyd, Quantum algorithm for linear systems of equations. Phys. Rev. Lett. **103**(15), 150502 (2009). https://doi.org/10.1103/PhysRevLett.103.150502
2. A.M. Childs, R. Kothari, R.D. Somma, Quantum algorithm for systems of linear equations with exponentially improved dependence on precision. SIAM J. Comput. **46**(6), 1920–1950 (2017). https://doi.org/10.1137/16M1087072
3. A. Gilyén, Y. Su, G.H. Low, N. Wiebe, Quantum singular value transformation and beyond: exponential improvements for quantum matrix arithmetics, in *Proceedings of the 51st Annual ACM SIGACT Symposium on Theory of Computing* (ACM, Phoenix, 2019), pp. 193–204. https://doi.org/10.1145/3313276.3316366
4. A. Ambainis, Variable time amplitude amplification and quantum algorithms for linear algebra problems, in *Leibniz International Proceedings in Informatics* (Schloss Dagstuhl - Leibniz-Zentrum für Informatik, 2012), pp. 636–647. https://doi.org/10.4230/LIPICS.STACS.2012.636
5. D. Orsucci, V. Dunjko, On solving classes of positive-definite quantum linear systems with quadratically improved runtime in the condition number. Quantum **5**, 573 (2021). https://doi.org/10.22331/q-2021-11-08-573
6. D. An, L. Lin, Quantum linear system solver based on time-optimal adiabatic quantum computing and quantum approximate optimization algorithm. ACM Trans. Quantum Comput. **3**(2), 1–28 (2022). https://doi.org/10.1145/3498331
7. B.D. Clader, B.C. Jacobs, C.R. Sprouse, Preconditioned quantum linear system algorithm. Phys. Rev. Lett. **110**(25), 250504 (2013). https://doi.org/10.1103/PhysRevLett.110.250504
8. J. Golden, D. O'Malley, H. Viswanathan, Quantum computing and preconditioners for hydrological linear systems. Sci. Rep. **12**(1), 22285 (2022). https://doi.org/10.1038/s41598-022-25727-9
9. C. Shao, H. Xiang, Quantum circulant preconditioner for a linear system of equations. Phys. Rev. A **98**(6), 062321 (2018). https://doi.org/10.1103/PhysRevA.98.062321
10. Y. Tong, D. An, N. Wiebe, L. Lin, Fast inversion, preconditioned quantum linear system solvers, fast Green's-function computation, and fast evaluation of matrix functions. Phys. Rev. A **104**(3), 032422 (2021). https://doi.org/10.1103/PhysRevA.104.032422
11. D.W. Berry, High-order quantum algorithm for solving linear differential equations. J. Phys. A: Math. Theor. **47**(10), 105301 (2014). https://doi.org/10.1088/1751-8113/47/10/105301
12. D.W. Berry, A.M. Childs, A. Ostrander, G. Wang, Quantum algorithm for linear differential equations with exponentially improved dependence on precision. Commun. Math. Phys. **356**(3), 1057–1081 (2017). https://doi.org/10.1007/s00220-017-3002-y
13. O.M. Raisuddin, S. De, qRLS: quantum relaxation for linear systems in finite element analysis. Eng. Comput. (2024). https://doi.org/10.1007/s00366-024-01975-3
14. A. Montanaro, S. Pallister, Quantum algorithms and the finite element method. Phys. Rev. A **93**(3), 032324 (2016). https://doi.org/10.1103/PhysRevA.93.032324
15. N. Linden, A. Montanaro, C. Shao, Quantum vs. classical algorithms for solving the heat equation. Commun. Math. Phys. **395**(2), 601–641 (2022). https://doi.org/10.1007/s00220-022-04442-6
16. D. An, J.-P. Liu, D. Wang, Q. Zhao, A theory of quantum differential equation solvers: limitations and fast-forwarding (2022). https://doi.org/10.48550/ARXIV.2211.05246

Quantum Linear System Algorithms: Iterative Methods

31

Although the QLSAs outlined in the previous chapter demonstrate an exponential speedup in N, even the optimal direct QLSA scales linearly with the condition number κ. The linear scaling in κ implies that exponential speedup is achievable only for systems with $\kappa = O(\text{poly}\log N)$. Practical problems of interest rarely exhibit such scaling. Classical approaches to large-scale linear systems often exploit prior knowledge through preconditioning or iterative solvers.

A quantum sparse approximate inverse (SPAI) preconditioner was proposed in [1] to resolve this issue. The approach produces the preconditioner by solving a least-squares problem on each row of the matrix. However, the SPAI preconditioner can be inefficient for many problems [2], and a complete implementation remains unavailable. A circulant matrix preconditioner has been proposed by Shao and Xiang [3], but its complexity depends on the condition number of the preconditioner $\kappa(M)$ and the product $\kappa\left(M^{-1}A\right)$, which in the worst case can satisfy $\kappa(M) \geq \kappa(A)$ and $\kappa(M), \kappa\left(M^{-1}A\right) \approx \sqrt{\kappa(A)}$. A Laplacian preconditioner has been proposed by Golden et al. [2] for hydrological subsurface flow to reduce the condition number to at most $O\left(\sqrt{N}\right)$. Tong et al. [4] presented a fast-inversion procedure to solve QLSPs of the form $(A + B)|x\rangle = |b\rangle$ where $\|A\| \ll \|B\|$, targeting applications such as single-particle Green's functions of quantum many-body systems.

A promising alternative is to develop iterative quantum algorithms. Optimal classical linear system solvers, such as the multigrid method, can achieve $O\left(N \log \frac{1}{\epsilon}\right)$ floating-point operations, independent of κ, or $O\left(\log \frac{1}{\epsilon}\right)$ matrix-vector multiplications. However, classical iterative linear system algorithms often require nonlinear operations (e.g., computing the L_2 norm in the conjugate-gradient method), which necessitates ancilla qubits and/or measurement in a quantum setting. As a result, many iterative methods are considered challenging for quantum

O. M. Raisuddin and S. De, *Quantum Computing for Engineers*,
https://doi.org/10.1007/978-3-032-03325-3_31

computers. Nevertheless, some iterative algorithms—such as Jacobi, Gauss–Seidel, and Richardson iterations—are composed solely of affine linear operations and are potentially more feasible for quantum implementation. In this chapter, we review recent progress in developing iterative quantum linear system algorithms.

We note that the variational quantum linear solver may be considered a hybrid quantum-classical "iterative" solver; however, we discuss it separately in Chap. 36, Notable Variational Algorithms: VQE, QAOA, and VQLS, alongside other variational algorithms.

We begin by defining the quantum linear system problem in the iterative setting as follows.

Definition (*Iterative Quantum Linear System Problem*) Let $A \in \mathbb{C}^{N \times N}$ be a given matrix and $b \in \mathbb{C}^N$ a target vector; given any initial vector $x^{(0)} \in \mathbb{C}^N$ such that $\left\|Ax^{(0)} - b\right\|_2 \leq \epsilon_0$ for some $\epsilon_0 > 0$, and given quantum oracle access to the entries of A and procedures for preparing the normalized quantum states $\left|x^{(0)}\right\rangle = \frac{\sum_i \left(x^{(0)}\right)_i |i\rangle}{\|x^{(0)}\|_2}$ and $|b\rangle = \frac{\sum_i b_i |i\rangle}{\|b\|_2}$, the goal is to output a quantum state $\left|x^{(l)}\right\rangle = \frac{\sum_i \left(x^{(l)}\right)_i |i\rangle}{\|x^{(l)}\|_2}$ for some l, such that $\left\|Ax^{(l)} - b\right\|_2 \leq \epsilon_l \leq \epsilon_0$.

For positive-definite linear systems, linear stationary iterations can be used. A general first-order linear stationary iteration may be defined as the affine linear operation:

$$x^{(l+1)} = \left(I - \tau_l C^{-1} A\right) x^{(l)} + \tau_l C^{-1} b$$

where C is a left-preconditioning matrix and τ_l is a parameter of the scheme. For Richardson iteration, set $C = I$ and $\tau_l = \omega$, giving

$$x^{(l+1)} = (I - \omega A)x^{(l)} + \omega b$$

Convergence requires the spectral radius $\rho(R) < 1$ where $R = (I - \omega A)$. This can be satisfied by choosing $\omega < \frac{1}{\lambda_{\max}(A)}$.

We can therefore rewrite this in terms of the block-linear system of equations framework in Chap. 24. As an example, for two iterations with two copy steps, we get the linear system:

$$\begin{bmatrix} I & & & & \\ -R & I & & & \\ & -R & I & & \\ & & -I & I & \\ & & & -I & I \end{bmatrix} \begin{bmatrix} x^{(0)} \\ x^{(1)} \\ x^{(2)} \\ x^{(3)} \\ x^{(4)} \end{bmatrix} = \begin{bmatrix} x^{(0)} \\ \tau b \\ \tau b \\ 0 \\ 0 \end{bmatrix}$$

For l iterations and c copies, we may rewrite the large block matrix as

$$M_{l,c} = \sum_{i=0}^{l+c} |i\rangle\langle i| \otimes I - \sum_{i=1}^{l} |i\rangle\langle i-1| \otimes R - \sum_{i=l+1}^{l+c} |i\rangle\langle i-1| \otimes I$$

Although this matrix is substantially larger than the original system, its condition number solely depends on the number of iterations l and copy steps. Furthermore, the number of copy steps can be chosen to be $c = l - 1$. With these choices, the condition number scales linearly with the number of iterations.

Lemma [5] $\forall\ M \in \mathbb{R}^{lN \times lN}$ *s.t.* $M_{ik} \in \mathbb{R}^{N \times N}$ *where* $M_{ii} = I, M_{i+1,i} = -R_i$, $M_{ji} = 0$ *and* $\forall\ i, j, k \in [1, l] \subset \mathbb{N}$ *where* $j \neq i, i+1$ *and* $\|R_i\| \leq 1\ \forall\ i$

$$\|M\| \leq 2, \left\|M^{-1}\right\| \leq l$$

$$\kappa_M \leq 2l$$

This technique has been analyzed by Raisuddin and De [5] *to arrive at the following result for an iterative QLSA as follows.*

Theorem *Let A be a symmetric positive-semidefinite matrix and* $M_{l,l-1}$ *denote the block matrix that encodes l linear stationary iterations for the system* $Ax = b$*, starting from the initial guess* $x^{(0)}$*. Then, for any* $\epsilon \leq \frac{1}{2}$*, there exists a quantum algorithm that prepares the normalized quantum state* $\left|x^{(l)}\right\rangle$ *corresponding to l linear stationary iterations with an overall complexity* $O\left(l \text{ poly} \log\left(\frac{lN}{\epsilon}\right)\right)$.

This is an exponential improvement over a classical computer in the problem size N. Naïve application of such iterative methods will require $l = O(\kappa)$ iterations, which matches the scaling of direct QLSAs. However, this algorithm is intended to be a building block for sophisticated iterative methods that leverage the structure of the problem and utilize linear stationary iterations as a subroutine, e.g., multigrid methods, domain decomposition, and block-preconditioners.

An iterative quantum linear system problem can be viewed as a gradient descent problem for the functional

$$f(x) = \frac{1}{2}x^T Ax - b^T x$$

Algorithms for gradient descent have been proposed by [6, 7]. In particular, Kerenidis and Prakash [6] presented a QRAM-based algorithm for gradient descent on linear systems, with complexity $O\left(\kappa^3\right)$ in the condition number and $O(1/\epsilon)$ in precision, along with other parameters.

Second-order linear stationary methods, such as the Chebyshev iteration, offer quadratic speedup for symmetric positive-definite systems, though no quantum algorithm is known for these methods. Raisuddin and De [8] developed a procedure for multigrid operations for finite element problems using block-matrix multiplication (see Chap. 26: Matrix-Vector Multiplications and Affine Linear Operations) to apply the sequence of affine linear operations. However, the success probability of the algorithm decreases with increasing problem size or precision demands.

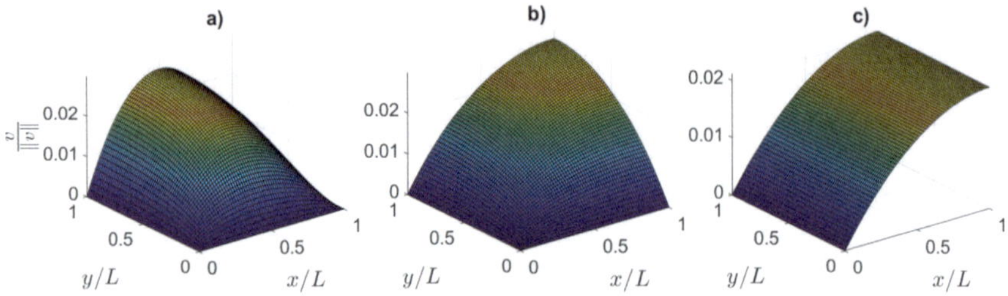

Fig. 31.1 Exact converged solutions obtained using classical solvers for the problem solved in this chapter

Below, we provide code for an iterative QLSA [5] implementing Richardson iterations for heat transfer problems in 1- and 2-dimensions with various boundary conditions; see Figs. 31.1 and 31.2 for convergence results.

```
#!/usr/bin/python3

import qiskit
from qiskit import *
import numpy as np
from scipy.io import loadmat
from scipy.io import savemat

print("Get phase angles from QSPPACK and store them as phi.mat")
print("Run this code by passing problem parameters in the following
format:")
print("python this_code.py problem_number num_qubits_each_
dimension num_iterations")

import sys
# print ('argument list', sys.argv)
# problem = int(sys.argv[1])
# n = int(sys.argv[2])
# l = int(sys.argv[3])

problem = 1
n = 2
l = 2

if problem == 1:
  d = 1
  NBCs = [[False, False]]
```

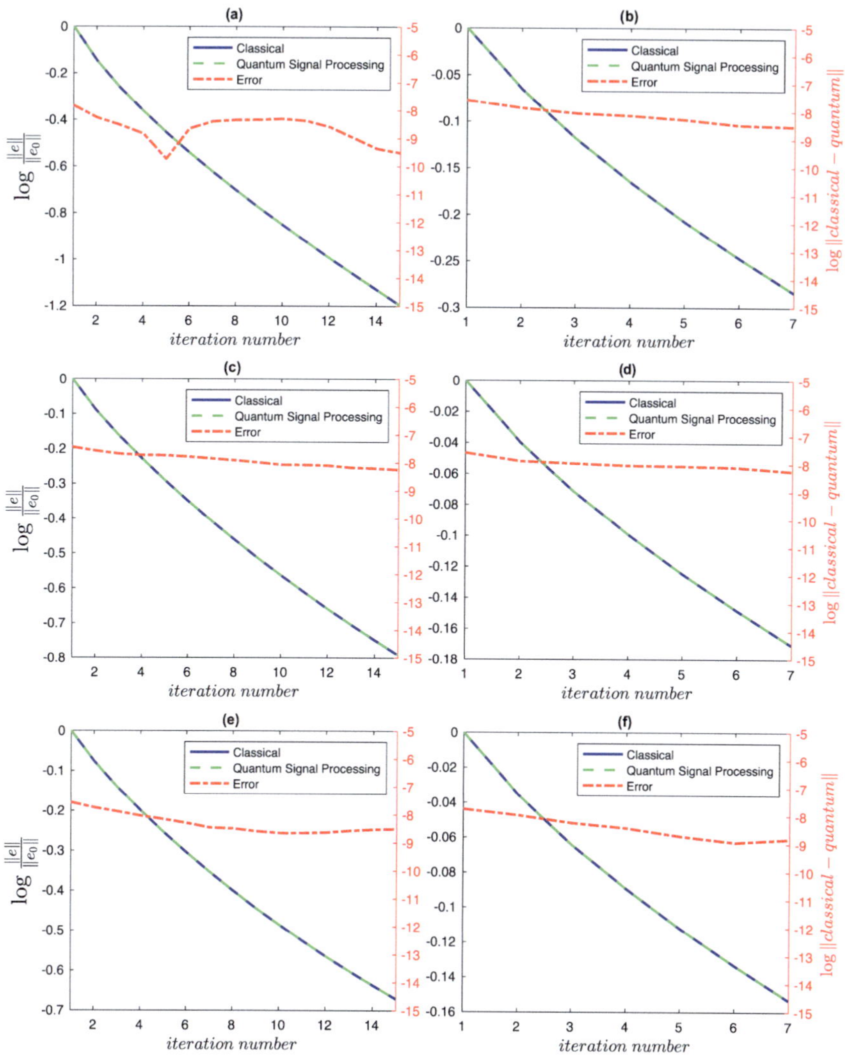

Fig. 31.2 Convergence of iterates produced classically and using a quantum iterative linear solver running on a simulator for the problem solved in this chapter

```
elif problem == 2:
    d = 1
    NBCs = [[False, True]]
elif problem == 3:
    d = 2
    NBCs = [[False, True],[False, False]]
elif problem == 4:
    d = 2
```

```
    NBCs = [[False, True],[False, True]]
elif problem == 5:
    d = 2
    NBCs = [[False, True],[True, True]]


print("problem number: {}".format(problem))
print("boundary conditions: {}".format(NBCs))
print("problem size: {}".format(n))
print("number of iterations {}".format(l))

# Load and prep angles
phi_angles = np.array( loadmat('phi_kappa_80_pts_8000_deg_
1999.mat') ).item()['phi_proc']
phase_angles = phi_angles.reshape(phi_angles.shape[0])

########### Functions for C ####################3
def C_i(i,register):
    n = register.size
    if i<1 or i>(n-1):
        print('WRONG VALUE FOR i !!!!')
        return
    Ci = QuantumCircuit(register, name='C_{}'.format(i))
    [Ci.cx(control_qubit=i, target_qubit=(i-j), ctrl_state='0') for
j in range(1,i+1)]
      Ci.append(qiskit.circuit.library.MCXGate(num_ctrl_qubits=i,
ctrl_state='0'*i),register[:i+1])
#               Ci.mcx(control_qubits=workRegister[:i-1],target_
qubit=workRegister[i-1])
    [Ci.cx(control_qubit=i, target_qubit=(i-j), ctrl_state='0') for
j in reversed(range(1,i+1))]
    return Ci

############### Functions for R ###############
def L1_d(register):
    n = register.size

    # Circuit that creates L1 unitary
    L1 = QuantumCircuit(n,name='L1')
    L1.x(0)

    return L1

def L2_d(register):
    n = register.size
```

```
    L2 = QuantumCircuit(register,name='L2')
    for j in range(1,n):
        L2 = L2.compose(C_i(j,register))

    return L2

def L3_d(register,NBC):
    n = register.size

    L3 = QuantumCircuit(register,name='L3')

    if not NBC[0]:
        L3.x(0)
        L3.h(0)
        L3.append(qiskit.circuit.library.MCXGate(num_ctrl_qubits=n-
1, ctrl_state='0'*(n-1)),register[1:]+[register[0]])
        L3.h(0)
        L3.x(0)

    if not NBC[1]:
        L3.h(n-1)
        L3.append(qiskit.circuit.library.MCXGate(num_ctrl_qubits=n-
1, ctrl_state='1'*(n-1)),register)
        L3.h(n-1)

    return L3

def L4_d(register,NBC):
    n = register.size

    L4 = QuantumCircuit(register,name='L4')

    # Apply -ve sign to the unitary
    L4.z(0)
    L4.x(0)
    L4.z(0)
    L4.x(0)

    return L4

def R_circuit(n,d,workRegisters,lcuRegister,NBCs,alphas):

    # Create the Prep Circuit
```

```
    prep = QuantumCircuit(lcuRegister,name='Prep')
    prep.prepare_state(alphas)

    allregisters = []
    allregisters.extend(workRegisters)
    allregisters.extend([lcuRegister])

  blockEncoded = QuantumCircuit(*workRegisters,lcuRegister,name='R')

    # Apply the PREP operation
    blockEncoded = blockEncoded.compose(prep,lcuRegister)

     # Apply the SELECT operation using controlled versions of the
circuits L1-L3
    # This needs to be done for each dimension!
    for i in range(d):
      if d>1:
        d_string = format(i, '0{}b'.format(d_size))
      else:
        d_string = ''

blockEncoded.append(L1_d(workRegisters[i]).control(num_ctrl_
qubits=(d_size+2) ,ctrl_state=d_string+'00'), lcuRegister[:] +
workRegisters[i][:])

blockEncoded.append(L2_d(workRegisters[i]).control(num_ctrl_
qubits=(d_size+2) ,ctrl_state=d_string+'01'), lcuRegister[:] +
workRegisters[i][:])

blockEncoded.append(L3_d(workRegisters[i],NBCs[i]).control(num_
ctrl_qubits=(d_size+2),ctrl_state=d_string+'10'),     lcuRegis-
ter[:] + workRegisters[i][:])

blockEncoded.append(L4_d(workRegisters[i],NBCs[i]).control(num_
ctrl_qubits=(d_size+2),ctrl_state=d_string+'11'),     lcuRegis-
ter[:] + workRegisters[i][:])

    # Apply the PREP+ operation
  blockEncoded = blockEncoded.compose(prep.inverse(),lcuRegister)

    return blockEncoded

######################## Functions for D ##########################
```

```
def L1_1(register):
    n = register.size

    # Circuit that creates L1 unitary
    L1 = QuantumCircuit(n,name='L1_1')
    L1.x(0)
    L1.z(0)

    return L1

def L1_2(register):
    n = register.size

    # Circuit that creates L1 unitary
    L1 = QuantumCircuit(n,name='L1_2')
    L1.z(0)
    L1.x(0)
    L1.z(0)

    return L1

def C_i(i,register):
    n = register.size
    if i<1 or i>(n-1):
        print('WRONG VALUE FOR i !!!!')
        return
    Ci = QuantumCircuit(register, name='C_{}'.format(i))
   [Ci.cx(control_qubit=i, target_qubit=(i-j), ctrl_state='0') for
j in range(1,i+1)]
      Ci.append(qiskit.circuit.library.MCXGate(num_ctrl_qubits=i,
ctrl_state='0'*i),register[:i+1])
   [Ci.cx(control_qubit=i, target_qubit=(i-j), ctrl_state='0') for
j in reversed(range(1,i+1))]
    return Ci

def L2_1(register):
    n = register.size

    L2 = QuantumCircuit(register,name='L2_1')
    # Apply -ve to sign to alternating bits
    L2.z(0)
    for j in range(1,n):
        L2 = L2.compose(C_i(j,register))
```

```
    return L2

def L2_2(register):
    n = register.size

    L2 = QuantumCircuit(register,name='L2_2')
    L2.h(n-1)
   L2.append(qiskit.circuit.library.MCXGate(num_ctrl_qubits=n-1,
ctrl_state='1'*(n-1)),register)
    L2.h(n-1)
    for j in range(1,n):
       L2 = L2.compose(C_i(j,register))
    L2.z(0)
    L2.x(0)
    L2.z(0)
    L2.x(0)

    return L2

def D_circuit(l,indexRegister,lcuRegister,alphas):

    # Create the Prep Circuit
    prep = QuantumCircuit(lcuRegister,name='Prep')
    prep.prepare_state(alphas)
    allregisters = []
    allregisters.extend([indexRegister])
    allregisters.extend([lcuRegister])

   blockEncoded = QuantumCircuit(indexRegister,lcuRegister,name='D')

    # Apply the PREP operation
    blockEncoded = blockEncoded.compose(prep,lcuRegister)

     # Apply the SELECT operation using controlled versions of the
circuits L1_i, L2_i
        blockEncoded.append(L1_1(indexRegister).control(num_ctrl_
qubits=(2) ,ctrl_state='00'), lcuRegister[:] + indexRegister[:])
        blockEncoded.append(L1_2(indexRegister).control(num_ctrl_
qubits=(2) ,ctrl_state='01'), lcuRegister[:] + indexRegister[:])
        blockEncoded.append(L2_1(indexRegister).control(num_ctrl_
qubits=(2) ,ctrl_state='10'), lcuRegister[:] + indexRegister[:])
        blockEncoded.append(L2_2(indexRegister).control(num_ctrl_
qubits=(2) ,ctrl_state='11'), lcuRegister[:] + indexRegister[:])
```

```
    # Apply the PREP+ operation
  blockEncoded = blockEncoded.compose(prep.inverse(),lcuRegister)

    return blockEncoded

################ Functions for QSP Rotation
###################################
def CR_phi_d_efficient(phi, _d, signalReg, lcuRegister_R, lcuReg-
ister_l, lcuRegister_q, circuit):
  BE_size = lcuRegister_R.size + lcuRegister_l.size + lcuRegister_
q.size
    ctrl_state = '0'*(BE_size)

        circuit.append(qiskit.circuit.library.MCXGate(num_ctrl_
qubits=BE_size, ctrl_state=ctrl_state), lcuRegister_R[:] +
lcuRegister_l[:] + lcuRegister_q[:] + signalReg[:])

    circuit.rz(2*phi, signalReg)
    circuit.z(signalReg)

        circuit.append(qiskit.circuit.library.MCXGate(num_ctrl_
qubits=BE_size, ctrl_state=ctrl_state), lcuRegister_R[:] +
lcuRegister_l[:] + lcuRegister_q[:] + signalReg[:])

    return

####################### Construct qRLS Circuit
###############################
d_size = int(np.log2(d))

workRegisters = [QuantumRegister(n,name='dim {}'.format(i)) for i
in range(d)]
indexRegister = QuantumRegister(1,name='index')

alphas_R = np.array([np.sqrt(1+0j), np.sqrt(1+0j),
np.sqrt(0.5+0j), np.sqrt(0.5+0j)]*d)
alphas_R = alphas_R/np.linalg.norm(alphas_R,2)
lcu_size_R = int(np.ceil(np.log2(alphas_R.size)))
lcuRegister_R = QuantumRegister(lcu_size_R,name='lcu_R')
R = R_circuit(n,d,workRegisters,lcuRegister_R,NBCs,alphas_R)

alphas_l = np.array([np.sqrt(0.5+0j), np.sqrt(0.5+0j),
np.sqrt(0.5+0j), np.sqrt(0.5+0j)])
alphas_l = alphas_l/np.linalg.norm(alphas_l,2)
```

```
lcu_size_l = int(np.ceil(np.log2(alphas_l.size)))
lcuRegister_l = QuantumRegister(lcu_size_l,name='lcu_l')
D = D_circuit(l,indexRegister,lcuRegister_l,alphas_l)

alphas_q = np.array([np.sqrt(1+0j), np.sqrt(3+0j)])
alphas_q = alphas_q/np.linalg.norm(alphas_q,2)
lcu_size_q = int(np.ceil(np.log2(alphas_q.size)))
lcuRegister_q = QuantumRegister(lcu_size_q,name='lcu_q')

R = R_circuit(n,d,workRegisters,lcuRegister_R,NBCs,alphas_R)
D = D_circuit(l,indexRegister,lcuRegister_l,alphas_l)
R = R.decompose(reps=6)
D = D.decompose(reps=6)

prep = QuantumCircuit(lcuRegister_q,name='Prep')
prep.prepare_state(alphas_q)
qRLS_circuit = QuantumCircuit(*workRegisters,indexRegister,lcuRegiste
R,lcuRegister_l,lcuRegister_q,name='qRLS')
qRLS_circuit = qRLS_circuit.compose(prep,lcuRegister_q)
qRLS_circuit.append(R.control(num_ctrl_qubits=(1)         ,ctrl_
state='1'), lcuRegister_q[:] + [_x for _xs in workRegisters for _x
in _xs] + lcuRegister_R[:])
qRLS_circuit.append(D.control(num_ctrl_qubits=(1)         ,ctrl_
state='1'),  lcuRegister_q[:] + indexRegister[:] + lcuRegister_
l[:] )
qRLS_circuit = qRLS_circuit.compose(prep.inverse(),lcuRegister_
q)

U_A = qRLS_circuit
U_A_i = U_A.inverse()

signalRegister = QuantumRegister(1,name='QSP signal')

QSP_circuit = QuantumCircuit(*workRegisters,indexRegister,lcuRegister_
R,lcuRegister_l,lcuRegister_q,signalRegister,name='QSP_
Solver')

# Prepare initial state
initial_state = np.ones(2**l)
initial_state[0] = 0
initial_state = initial_state/np.linalg.norm(initial_state,2)

QSP_circuit.append(qiskit.circuit.library.StatePreparation(initial_
state),indexRegister)
```

```
for _i in workRegisters:
  QSP_circuit.h(_i)

#######################         Start         QSP         Sequence
#######################
# First thing is to Hadamard the signal qubit since we want Re(P(A))
QSP_circuit.h(signalRegister)

for _d, phi in reversed( list( enumerate( phase_angles[:]))):
        CR_phi_d_efficient(phi,_d,signalRegister,lcuRegister_
R,lcuRegister_l,lcuRegister_q,QSP_circuit)
  if _d>(0):
    if _d%2:
      for _ci in U_A_i.data:
        QSP_circuit.append(_ci)
    else:
      for _ci in U_A.data:
        QSP_circuit.append(_ci)

# Apply the final Hadamard gate
Q SP_circuit.h(signalRegister)

####################### Simulate Circuit #######################
print('running simulation')
from qiskit_aer import Aer
device = 'CPU'
backend       =       Aer.get_backend('statevector_simulator',
device=device, precision='double')

print('transpiling circuit')
transpiled_QSP_circuit          =          qiskit.transpile(QSP_
circuit.decompose(reps=3))

print('completed transpilation, starting job')
result = backend.run(transpiled_QSP_circuit,shots=0).result()

statevector = result.get_statevector()
final_output = np.array(statevector)[0:(2**(1)*2**(n*d))]
iterates = final_output/np.linalg.norm(final_output,2)

#######################     Calculate     classical     iterates
#######################
R = [np.zeros((2**n,2**n)) for _i in NBCs]
for _R in R:
```

```
  i, j = np.indices(_R.shape)
  _R[i==j-1] = 0.5
  _R[i==j+1] = 0.5
for _i,_R in zip(NBCs,R):
  if _i[0]:
    _R[0,0] = 0.5
  if _i[1]:
    _R[-1,-1] = 0.5
R_full = np.zeros((2**(n*d),2**(n*d)))
for _i,_R in enumerate(reversed(R)):
  if _i == 0:
    _Ri = _R
  else:
    _Ri = np.eye(2**n)
  for _j in range(1,d):
    if _j==_i:
      _Ri = np.kron(_Ri,_R)
    else:
      _Ri = np.kron(_Ri,np.eye(2**n))
  R_full += _Ri
R_full = R_full/d

classical_iterates = [np.zeros(R_full.shape[0]) for _i in
range(2**l)]
f = np.ones(R_full.shape[0])
for _i in range(2**l - 1):
      classical_iterates[_i+1] = np.matmul(R_full,classical_
iterates[_i]) + f
classical_iterates = np.concatenate(classical_iterates)
classical_iterates = classical_iterates/np.linalg.norm(classical_
iterates,2)

######################## Calculate the exact solution
########################
A = [np.zeros((2**n,2**n)) for _i in NBCs]
for _A in A:
  i, j = np.indices(_A.shape)
  _A[i==j-1] = -1
  _A[i==j+1] = -1
  _A[i==j] = 2
for _i,_A in zip(NBCs,A):
  if _i[0]:
    _A[0,0] = 1
  if _i[1]:
    _A[-1,-1] = 1
```

```
A_full = np.zeros((2**(n*d),2**(n*d)))
for _i,_A in enumerate(reversed(A)):
  if _i == 0:
    _Ai = _A
  else:
    _Ai = np.eye(2**n)
  for _j in range(1,d):
    if _j==_i:
      _Ai = np.kron(_Ai,_A)
    else:
      _Ai = np.kron(_Ai,np.eye(2**n))
  A_full += _Ai
f = np.ones(R_full.shape[0])
exact_sol = np.linalg.solve(A_full,f)
exact_sol = exact_sol/np.linalg.norm(exact_sol,2)

#######################    Calculate    the    iterate    errors
#######################
quantum_convergence = []
for _i in range(1,2**l):
            quantum_convergence.append(   np.linalg.norm(exact_
sol        -        iterates[(2**(n*d))*(_i):(2**(n*d))*(_i+1)]/
np.linalg.norm(iterates[(2**(n*d))*(_i):(2**(n*d))*(_i+1)],2))
)

classical_convergence = []
for _i in range(1,2**l):
          classical_convergence.append(   np.linalg.norm(exact_
sol      -      classical_iterates[(2**(n*d))*(_i):(2**(n*d))*(_
i+1)]/np.linalg.norm(classical_iterates[(2**(n*d))*(_
i):(2**(n*d))*(_i+1)],2)) )

#######################     Calculate     the     QSP     errors
#######################
QSP_errors_full = []
for _i in range(2**l):
        QSP_errors_full.append( np.linalg.norm(  (  classical_
iterates[(2**(n*d))*(_i):(2**(n*d))*(_i+1)])    -    iterates[_
i*(2**(n*d)):(_i+1)*(2**(n*d))], 2 ) )

#######################     Save     all     useful     variables
#######################
mdic   =   {"dimensions":d,   "n":n,   "l":l,   "BCs":NBCs,   "raw_
statevector":   final_output,   "classical_solution":classical_
```

```
iterates, "QSP_solution":iterates, "QSP_Errors":QSP_errors_
full, "classical":classical_convergence, "quantum":quantum_
convergence}
savemat("output_problem_{}_n_{}_l_{}.mat".format(problem,n,l),
mdic)

print("completed simulation with parameters:")
print("problem number: {}".format(problem))
print("boundary conditions: {}".format(NBCs))
print("problem size: {}".format(n))
print("number of iterations {}".format(l))

print('\n\nsaved output\n\n')
```

In summary, iterative quantum linear system algorithms are a recent and active area of quantum computing research. While methods such as the qRLS demonstrate the feasibility of quantum approaches for solving linear systems, their applicability is presently limited to symmetric positive-definite matrices due to convergence requirements and the structure of quantum algorithmic primitives. Key challenges remain in extending these techniques to more general matrix classes, improving convergence for ill-conditioned or indefinite systems, and developing robust quantum preconditioners. Addressing these limitations will be essential for advancing the practical impact and applicability of iterative quantum algorithms in scientific and engineering domains.

References

1. B.D. Clader, B.C. Jacobs, C.R. Sprouse, Preconditioned quantum linear system algorithm. Phys. Rev. Lett. **110**(25), 250504 (2013). https://doi.org/10.1103/PhysRevLett.110.250504
2. J. Golden, D. O'Malley, H. Viswanathan, Quantum computing and preconditioners for hydrological linear systems. Sci. Rep. **12**(1), 22285 (2022). https://doi.org/10.1038/s41598-022-25727-9
3. C. Shao, H. Xiang, Quantum circulant preconditioner for a linear system of equations. Phys. Rev. A **98**(6), 062321 (2018). https://doi.org/10.1103/PhysRevA.98.062321
4. Y. Tong, D. An, N. Wiebe, L. Lin, Fast inversion, preconditioned quantum linear system solvers, fast Green's-function computation, and fast evaluation of matrix functions. Phys. Rev. A **104**(3), 032422 (2021). https://doi.org/10.1103/PhysRevA.104.032422
5. O.M. Raisuddin, S. De, qRLS: quantum relaxation for linear systems in finite element analysis. Eng. Comput. (2024). https://doi.org/10.1007/s00366-024-01975-3
6. I. Kerenidis, A. Prakash, Quantum gradient descent for linear systems and least squares. Phys. Rev. A **101**(2), 022316 (2020). https://doi.org/10.1103/PhysRevA.101.022316
7. P. Rebentrost, M. Schuld, L. Wossnig, F. Petruccione, S. Lloyd, Quantum gradient descent and Newton's method for constrained polynomial optimization. New J. Phys. **21**(7), 073023 (2019). https://doi.org/10.1088/1367-2630/ab2a9e
8. O.M. Raisuddin, S. De, Quantum multigrid algorithm for finite element problems (2024). https://doi.org/10.48550/ARXIV.2404.07466

Quantum Ordinary Differential Equation Algorithms: Block-Matrix Algorithms

32

Systems of ordinary differential equations (ODEs) frequently arise in scientific and engineering computations. For many problems of practical interest, the systems may be high-dimensional and large enough to be computationally expensive, to the extent that they become large enough to be computationally prohibitive or intractable. The capability of quantum computers to process vectors in exponentially large spaces is a promising solution to alleviate computational resource concerns for the numerical solution of large-scale differential equation problems. In this chapter we discuss the block-matrix approach for quantum ordinary differential equation algorithms. In the next chapter, the more nascent time-marching approach is discussed.

For inhomogeneous linear systems of ODEs, the first efficient quantum algorithm was developed by Berry [1] by using a linear multistep method for time discretization. The recursive relation for discrete time-stepping is encoded as a block-linear system, as discussed in Chap. 26: Matrix-Vector Multiplications and Affine Linear Operations. The scaling was subsequently improved in [2] to achieve quasi-linear scaling in t and exponentially improved precision by using a Taylor series time discretization instead of a linear multistep formula. A spectral time discretization approach was provided by Childs and Liu [3], which can also accommodate the case of time-dependent $A(t)$. We note that all these algorithms require that A have non-positive real parts of its eigenvalues, i.e., only decaying solutions in the 2-norm are tractable. The results of [1–3] were further improved and generalized by Krovi [4], with exponentially improved bounds on error for ill-conditioned diagonalizable linear ODE systems, non-diagonalizable systems of ODEs, and also provided an improved version of the algorithm provided by Liu et al. [5] for Carleman linearized nonlinear PDEs. An et al. [6] developed a theory for the overhead in the quantum complexity of solving homogeneous ODEs, identifying sources of

O. M. Raisuddin and S. De, *Quantum Computing for Engineers*,
https://doi.org/10.1007/978-3-032-03325-3_32

"non-quantumness" that increase overhead compared to quantum dynamics simulation. Specifically, non-unique real parts of the eigenvalues can lead to exponential overhead in the worst case, while the non-orthogonality of eigenvectors, measured by $\mu(A) = \left\|AA^\dagger - A^\dagger A\right\|^{1/2}$, incurs a linear overhead in $\mu(A)$.

For simplicity, we outline the method of [2] as an example, solving a first-order time-independent initial value problem for the ODE system

$$\frac{\mathrm{d}x}{\mathrm{d}t} = Ax + b$$

with initial condition $x(0)$, where A and b are time-independent.

This ODE system has an exact solution

$$x(t) = e^{At}x(0) + \left(e^{At} - I\right)A^{-1}b$$

e^{At} and $\left(e^{At} - I\right)A^{-1}$ are approximated using Taylor series expansions with $k+1$ and k terms, respectively:

$$e(z) \approx T_k(z) = \sum_{j=0}^{k} \frac{z^j}{j!}, \quad z^{-1} \approx S_k(z) = \sum_{j=1}^{k} \frac{z^{j-1}}{j!}$$

A time step $h \leq \frac{1}{\|A\|}$ is chosen, yielding the recurrence

$$x_t = T_k(Ah)x_{t-1} + S_k(Ah)b$$

where $x_0 = x(0)$ and $x_t \approx x(t)$ such that $\left\| \frac{x_t}{\|x_t\|} - \frac{x(t)}{\|x(t)\|} \right\| \leq \epsilon$. This recurrence relation involves powers of Ah and their sums, making the block-linear systems slightly more complex for this problem, compared to the much simpler sequence of linear operations discussed in Chap. 26: Matrix-Vector Multiplications and Affine Linear Operations. p copy steps are also appended to boost the success probability, which are also discussed in Chap. 26: Matrix-Vector Multiplications and Affine Linear Operations. The block-linear system is of the form:

$$M_{A,h,k,m,p}|x\rangle = |\text{init}\rangle$$

where

$$M_{A,h,k,m,p} = \begin{pmatrix} I & & & & & & & & & & & & & \\ -Ah & I & & & & & & & & & & & & \\ & -\frac{Ah}{2} & I & & & & & & & & & & & \\ & & \ddots & \ddots & & & & & & & & & & \\ & & & -\frac{Ah}{k} & I & & & & & & & & & \\ -I & -I & \cdots & -I & -I & I & & & & & & & & \\ & & & & & -Ah & I & & & & & & & \\ & & & & & & \ddots & \ddots & & & & & & \\ & & & & & & & -\frac{Ah}{k} & I & & & & & \\ & & & & & -I & \cdots & -I & -I & I & & & & \\ & & & & & & & & & -I & I & & & \\ & & & & & & & & & & \ddots & \ddots & & \\ & & & & & & & & & & & -I & I \end{pmatrix}$$

$$|x\rangle = \sum_{i=0}^{m-1} \sum_{j=0}^{k} |i(k+1)+j\rangle \left|x_{i,j}\right\rangle + \sum_{j=0}^{p} |m(k+1)+j\rangle \left|x_{m,j}\right\rangle$$

$$= \begin{pmatrix} x_{0,0} \\ x_{0,1} \\ x_{0,2} \\ \vdots \\ x_{0,k} \\ x_{1,0} \\ x_{1,1} \\ \vdots \\ x_{m-1,k} \\ x_{m,0} \\ x_{m,1} \\ \vdots \\ x_{m,p} \end{pmatrix} = \begin{pmatrix} x_0 \\ Ahx_0 \\ (Ah)^2/2x_0 \\ \vdots \\ (Ah)^k/k!x_0 \\ x_1 \\ Ahx_1 \\ \vdots \\ (Ah)^k/k!x_{m-1} \\ x_m \\ x_m \\ \vdots \\ x_m \end{pmatrix}$$

$$|\text{init}\rangle = |0\rangle|x_0\rangle + h\sum_{i=0}^{m-1}|i(k+1)+1\rangle|b\rangle = \begin{pmatrix} x_0 \\ hb \\ 0 \\ \vdots \\ 0 \\ 0 \\ hb \\ \vdots \\ 0 \\ 0 \\ 0 \\ \vdots \\ 0 \end{pmatrix}$$

where normalization of $|x\rangle$ and $|\text{init}\rangle$ is implied.

Solving this linear system yields a quantum state $|x\rangle$ encoding all the time steps (and intermediate vectors). Finally, measuring the first register $|i\rangle$ such that $i \geq m(k+1)+j$ projects onto the desired solution.

This algorithm consists of the following three major steps:

1. Prepare $|\text{init}\rangle$
2. Solve linear system $M_{A,h,k,m,p}$
3. Measure first register $|i\rangle$
4. Restart if $i < m(k+1)+j$.

The overall complexity [2] is

$$\mathcal{O}(\kappa_A g T \|A\| \text{poly}\log(\kappa_A g \beta T \|A\|/\epsilon))$$

where κ_A is the condition number of A, $g = \max_{t\in[0,mh]} \frac{\||x(t)\rangle\|}{\||x(mh)\rangle\|}$, and $\beta = \frac{\||x_0\rangle\| + T\||b\rangle\|}{\||x(T)\rangle\|}$.

The method uses $(1, m, 0)$ block encoding of U_A, and state preparation unitaries U_{x_0} and U_b.

The no-fast-forwarding theorem forbids sublinear scaling in time for general problems [2]. While [6] provides lower bounds on the complexity of solving any general linear systems of ODEs, they also identify specific cases for which the solution can be fast-forwarded, i.e., solved with sublinear scaling in time. Bounded negative-semidefinite linear ODE problems with square-root access to A or negative-definite problems can achieve a quadratic speedup in T, while problems for which the Eigen decomposition is known can achieve an exponential speedup.

An algorithm with a complete circuit description and experimental results for a system of inhomogeneous ODEs of size 4×4 was given by Xin et al. [7] for unitary A.

Speedups for ODEs can have a great impact on engineering and scientific computation. As an example, the N-body problem involves numerically solving a system of ODEs in $6N$ dimensions for time t. Compared to a classical complexity of $O(t)$ and $O(N)$ for solution time and memory resources, a quantum computer could potentially require $O(t)$ time and $O(\log(N))$ qubits.

Hyperbolic and parabolic partial differential equations, ubiquitous in engineering, physics, finance, medicine, and many other applications, can be transformed into systems of ODEs using a discretization in space. We discuss partial differential equations in Chap. 34: Quantum Partial Differential Equation Algorithms.

The Hamiltonian simulation problem, discussed in detail in Chap. 28, Hamiltonian Simulation Techniques, for quantum dynamics of closed quantum systems seeks a solution of the Schrodinger equation, which is a homogeneous system of first-order ODEs. The first quantum algorithm for this problem was proposed by Lloyd [8] using the Trotter method. The complexity of this method was later improved to a query complexity of $O\big(d^2\|H\|_{\max}t\log\big(d^2\|H\|_{\max}/\epsilon\big)\big)$ by Berry et al. [9] by using a linear combination of unitaries arising from a truncated Taylor series. The optimal algorithm for Hamiltonian simulation using quantum signal processing was later provided by Low and Chuang [10] with a query complexity of $O\big(td\|H\|_{\max} + \log\big(\frac{1}{\epsilon}\big)/\log\log\big(\frac{1}{\epsilon}\big)\big)$ using d-sparse oracles or $O\big(t\|H\| + +\log\big(\frac{1}{\epsilon}\big)/\log\log\big(\frac{1}{\epsilon}\big)\big)$ using block-encoded oracles.

Various algorithms have been presented for nonlinear ODE problems. The first quantum algorithm for nonlinear differential equations (also the first quantum algorithm for any differential equation) was proposed by Leyton [11]. Leyton's algorithm proposed solving a nonlinear system of ordinary differential equations using the Euler method and requires multiple copies of the initial condition, leading to an exponentially increasing cost in time. The algorithm requires the preparation of multiple copies of a quantum state to effect nonlinear transformations of amplitude-encoded states. While block-encoding enables nonlinear transformations of the singular values of a matrix A, it does not allow nonlinear transformations of quantum state amplitudes as inputs to nonlinear functions.

Liu et al. [5] has provided an algorithm for dissipative nonlinear differential equations, specifically the n-dimensional quadratic ODE initial value problem, using the Carleman linearization technique. Xue et al. [12] applied homotopic perturbation methods for exponentially improved precision for the homogeneous version of the algorithm in [5] with an orthogonal linear term. However, Krovi [4] notes that the problem considered by Xue et al. [12] is limited by exponentially decaying solutions and suffers from the post-selection problem. Joseph [13] explores the transformation of phase space to an equivalent Schrödinger equation using the Koopman–von Neumann formulation. Berry and Costa [14] addresses time-dependent problems using the Dyson series, and the spectral method in [3] can also handle time-dependent $A(t)$. Lloyd et al. [15] proposed an algorithm for nonlinear ODEs using forward Euler discretization for short time intervals;

this also requires multiple copies of the initial state, but the increase in cost is quadratic rather than exponential as in [11]. A time-marching strategy for nonlinear ODEs has been suggested by [16, 17]. All quantum algorithms for nonlinear ODEs require the system to be dissipative.

Engel et al. [18] provides an overview of techniques for mapping nonlinear systems to infinite-dimensional linear systems using Carleman embedding and truncating to finite-dimensional systems for solution on quantum computers but does not provide a concrete algorithm. Joseph [13] also provides an overview of the Koopman–von Neumann method for a potential quadratic speedup in specific cases.

References

1. D.W. Berry, High-order quantum algorithm for solving linear differential equations. J. Phys. A: Math. Theor. **47**(10), 105301 (2014). https://doi.org/10.1088/1751-8113/47/10/105301
2. D.W. Berry, A.M. Childs, A. Ostrander, G. Wang, Quantum algorithm for linear differential equations with exponentially improved dependence on precision. Commun. Math. Phys. **356**(3), 1057–1081 (2017). https://doi.org/10.1007/s00220-017-3002-y
3. A.M. Childs, J.-P. Liu, Quantum spectral methods for differential equations. Commun. Math. Phys. **375**(2), 1427–1457 (2020). https://doi.org/10.1007/s00220-020-03699-z
4. H. Krovi, Improved quantum algorithms for linear and nonlinear differential equations. Quantum **7**, 913 (2023). https://doi.org/10.22331/q-2023-02-02-913
5. J.-P. Liu, H.Ø. Kolden, H.K. Krovi, N.F. Loureiro, K. Trivisa, A.M. Childs, Efficient quantum algorithm for dissipative nonlinear differential equations. Proc. Natl. Acad. Sci. U.S.A. **118**(35), e2026805118 (2021). https://doi.org/10.1073/pnas.2026805118
6. D. An, J.-P. Liu, D. Wang, Q. Zhao, A theory of quantum differential equation solvers: limitations and fast-forwarding (2022). https://doi.org/10.48550/ARXIV.2211.05246
7. T. Xin et al., Quantum algorithm for solving linear differential equations: theory and experiment. Phys. Rev. A **101**(3), 032307 (2020). https://doi.org/10.1103/PhysRevA.101.032307
8. S. Lloyd, Universal quantum simulators. Science **273**(5278), 1073–1078 (1996). https://doi.org/10.1126/science.273.5278.1073
9. D.W. Berry, A.M. Childs, R. Cleve, R. Kothari, R.D. Somma, Simulating Hamiltonian dynamics with a truncated Taylor series. Phys. Rev. Lett. **114**(9), 090502 (2015). https://doi.org/10.1103/PhysRevLett.114.090502
10. G.H. Low, I.L. Chuang, Optimal Hamiltonian simulation by quantum signal processing. Phys. Rev. Lett. **118**(1), 010501 (2017). https://doi.org/10.1103/PhysRevLett.118.010501
11. S.K. Leyton, T.J. Osborne, A quantum algorithm to solve nonlinear differential equations (2008). https://doi.org/10.48550/ARXIV.0812.4423
12. C. Xue, Y.-C. Wu, G.-P. Guo, Quantum homotopy perturbation method for nonlinear dissipative ordinary differential equations. New J. Phys. **23**(12), 123035 (2021). https://doi.org/10.1088/1367-2630/ac3eff
13. I. Joseph, Koopman–von Neumann approach to quantum simulation of nonlinear classical dynamics. Phys. Rev. Res. **2**(4), 043102 (2020). https://doi.org/10.1103/PhysRevResearch.2.043102
14. D.W. Berry, P.C.S. Costa, Quantum algorithm for time-dependent differential equations using Dyson series (2022). https://doi.org/10.48550/ARXIV.2212.03544
15. S. Lloyd et al., Quantum algorithm for nonlinear differential equations (2020). https://doi.org/10.48550/ARXIV.2011.06571
16. V. Buzek, M. Hillery, Quantum cloning. Phys. World **14**(11), 25–30 (2001). https://doi.org/10.1088/2058-7058/14/11/28

17. G.H. Low, I.L. Chuang, Hamiltonian simulation by qubitization. Quantum **3**, 163 (2019). https://doi.org/10.22331/q-2019-07-12-163
18. A. Engel, G. Smith, S.E. Parker, Linear embedding of nonlinear dynamical systems and prospects for efficient quantum algorithms. Phys. Plasmas **28**(6), 062305 (2021). https://doi.org/10.1063/5.0040313

Quantum Ordinary Differential Equation Algorithms: Time-Marching Algorithms

33

Recent work has demonstrated methods to circumvent the construction of a large linear system by using a time-marching strategy to propagate the solution forward in time [1, 2]. However, these techniques are currently sub-optimal and scale quadratically in t.

A time-marching algorithm can be described as a procedure that integrates differential equations one time step at a time, following the sequence

$$|\psi_{l-1}\rangle \rightarrow |\psi_l\rangle \rightarrow |\psi_{l+1}\rangle$$

For time-independent linear first-order ordinary differential equations (ODEs) with unitary dynamics, this can be accomplished in a relatively straightforward manner using the Trotter method. However, for higher precision and non-unitary dynamics, additional considerations are needed. The first quantum time-marching method was proposed by Fang et al. [2] for time-dependent ODEs, utilizing the Dyson series.

In this chapter, we will focus on challenges associated with the time-marching approach and techniques to address them. Consider a time-independent first-order homogeneous ODE system

$$\frac{\mathrm{d}x}{\mathrm{d}t} = Ax$$

Using a forward Euler discretization with time steps Δt, we obtain

$$x_t = (I + A\Delta t)x_{t-1}$$

O. M. Raisuddin and S. De, *Quantum Computing for Engineers*,
https://doi.org/10.1007/978-3-032-03325-3_33

Let's assume access to $I + A\Delta t$ through a $(\alpha\|I + A\Delta t\|, m, 0)$ block encoding. We may use this block encoding for each time step

$$\begin{pmatrix} \frac{I+A\Delta t}{\alpha\|I+A\Delta t\|} & * \\ * & * \end{pmatrix}\left(|x_{t-1}\rangle \otimes |0\rangle^{\otimes m}\right) = \frac{1}{\alpha\|I + A\Delta t\|}|x_t\rangle \otimes |0\rangle^{\otimes m} + |\perp\rangle$$

where successful measurement of all ancilla qubits in the $|0\rangle$ state yields $|x_t\rangle$ with a success probability of

$$p\left(|0\rangle^{\otimes m}\right) = \frac{1}{\alpha^2\|I + A\Delta t\|^2}\frac{\|x_t\|^2}{\|x_{t-1}\|^2}$$

Given $|x_0\rangle$, applying this process $L = \frac{T}{\Delta t}$ times yields $|x_T\rangle$ with an overall success probability

$$\frac{1}{\alpha^{2L}\|I + AT/L\|^{2L}}\frac{\|x_T\|^2}{\|x_0\|^2} \approx \frac{1}{\alpha^L}\frac{1}{\left\|e^{AT}\right\|^2}\frac{\|x_T\|^2}{\|x_0\|^2}$$

Note that this approach requires $O(mL)$ ancilla qubits. The factor $\frac{1}{\alpha^L}$ reflects the excessive subnormalization of the block-encoding. In the optimal case, $\alpha = 1$. The factors $\frac{1}{\|I+AT/L\|^{2L}}$ and $\frac{\||x_T\rangle\|^2}{\||x_0\rangle\|^2}$ arise from the block-encoding and the dynamics of the problem itself.

The excessive subnormalization arising due to α can be mitigated by applying the Uniform Singular Value Amplification (USVA) procedure. As discussed in Chap. 26, Matrix-Vector Multiplications and Affine Linear Operations, the underlying idea of USVA is to use the Quantum Singular Value Transform to apply a polynomial approximating $f(x) \approx g(x) = \frac{\alpha}{1-\delta}x$ over the interval $x \in \left[-\frac{1}{\alpha}, \frac{1}{\alpha}\right]$ as shown in Fig. 33.1. Since $f(x)$ approximates $g(x)$ with a polynomial, this method introduces errors.

By choosing $\delta = \frac{1}{L}$, the success probability over $L > 1$ time steps can be bounded below by

$$\frac{1}{(1-\delta)^L}\frac{1}{\|I + AT/L\|^{2L}}\frac{\|x_T\|^2}{\|x_0\|^2}$$

Using the inequality $(1-\delta)^L \geq e^{-\frac{\delta L}{1-\delta}}$, which for $\delta = \frac{1}{L}$ gives $(1-\delta)^L = (1-1/L)^L \geq e^{-1}$, we see that this term is bounded below by a constant, i.e., $\Omega(1)$. Therefore, the overall success probability is $\mathcal{O}\left(\frac{1}{\|I+AT/L\|^{2L}}\frac{\|x_T\|^2}{\|x_0\|^2}\right)$.

As discussed in Chap. 26, Matrix-Vector Multiplications and Affine Linear Operations, we may use a compression gadget to reduce the number of ancilla qubits. In the naïve approach for implementing a time-marching algorithm, one must dedicate $O(m)$ ancilla qubits for each block encoding, implementing a time

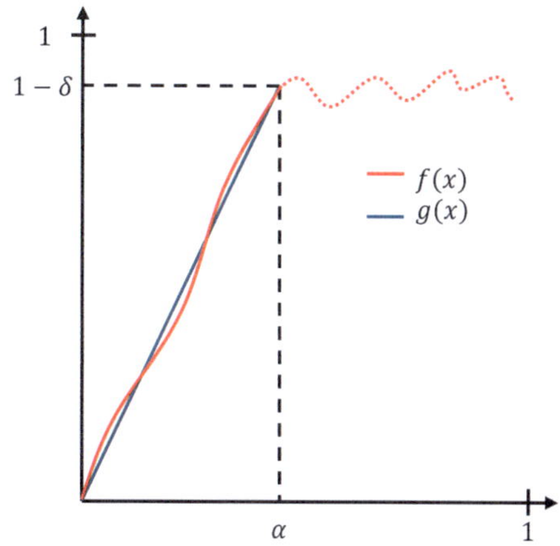

Fig. 33.1 Polynomial fit for USVA

step. Furthermore, the USVA procedure requires one more ancilla qubit. These ancilla requirements may be reduced using a compression gadget to $O(m + \log L)$.

Using these techniques, one may develop an algorithm for homogeneous time-dependent linear systems of differential equations based on the Dyson series approach

$$\frac{\mathrm{d}}{\mathrm{d}t}|\psi(t)\rangle = A(t)|\psi(t)\rangle$$

For full analysis and complexity of this algorithm, see [2].

It is important to note that this time-marching approach is not known to be optimal, as its complexity scales quadratically in t, compared to the linear scaling achievable by block-matrix methods. Furthermore, existing algorithms have been developed primarily for homogeneous systems, though the method of variation of parameters has been proposed for inhomogeneous cases. Whether time-marching algorithms for ODEs can match the efficiency of block-matrix methods remains an open research question.

References

1. D. An, J.-P. Liu, D. Wang, Q. Zhao, A theory of quantum differential equation solvers: limitations and fast-forwarding (2022). https://doi.org/10.48550/ARXIV.2211.05246
2. D. Fang, L. Lin, Y. Tong, Time-marching based quantum solvers for time-dependent linear differential equations. Quantum **7**, 955 (2023). https://doi.org/10.22331/q-2023-03-20-955

Quantum Partial Differential Equation Algorithms

34

Partial differential equations (PDEs) are central to modeling physical phenomena across science and engineering. PDEs of practical interest can commonly be classified as elliptic, parabolic, and hyperbolic systems. PDEs are solved numerically on classical computers using techniques such as the finite element, finite difference, or finite volume methods, which reduce the PDEs to a discretized system of equations. These equations are solved using either direct or iterative methods. Iterative solvers scale better than direct methods and are preferred for large-scale problems with steady-state solutions or transient problems with implicit time marching.

Quantum computers can be used to accelerate the solution of linear systems of equations arising from PDE discretization through quantum linear system algorithms (QLSA). However, for large discretizations, the linear system can become increasingly ill-conditioned. As an example, the condition number of a linear system arising from a finite element discretization can scale as $O(N^2)$, negating the exponential speedup provided by direct QLSAs when compared to classical iterative solvers. Alternatively, some PDEs can be mapped to either the Schrödinger equation or to systems of ODEs, enabling the use of quantum Hamiltonian simulation or ODE solvers that can yield exponential speedups with respect to the number of unknowns. However, such approaches are limited to structured grids on rectangular domains.

Several quantum algorithms have been proposed for solving PDEs using different discretization techniques: finite element [1, 2], finite volume [3], finite difference [4–6], and spectral methods [7]. These approaches include (i) directly solving the discretized linear system via QLSAs [1]; (ii) employing Hamiltonian simulation to extract eigenvalues [8, 9]; (iii) mapping the PDE to the Schrödinger equation and evolving in time using Hamiltonian simulation [10, 11]; and (iv) quantum ODE algorithms to evolve spatially discretized evolutionary PDEs [12]. We summarize these approaches below.

O. M. Raisuddin and S. De, *Quantum Computing for Engineers*,
https://doi.org/10.1007/978-3-032-03325-3_34

Clader et al. [1] introduces the use of QLSAs to solve the linear system arising from a finite element discretization with preconditioning using a sparse approximate-inverse preconditioner for exponential speedup. This approach was further investigated by Montanaro and Pallister [2], who points out that when the cost of reading out the properties of the solution is included, the speedup is polynomial, with the speedup increasing for higher dimensional problems. However, quantum circuits or procedures to implement the preconditioner were not provided.

For the Poisson equation with Dirichlet boundary conditions on rectangular grids, Cao et al. [9] proposed an algorithm with linear scaling in spatial dimension d and polylogarithmic scaling in $1/\epsilon$, utilizing the finite difference method to discretize the Laplacian operator on a unit cube and using Hamiltonian simulation of the discretized operator. Due to the geometry of the domain, the Laplacian operator in d dimensions can be expressed as the Kronecker sum

$$A = L_h \otimes I \otimes \cdots \otimes I + I \otimes L_h \otimes \cdots \otimes I + \cdots + I \otimes I \otimes \cdots \otimes L_h$$

where L_h is the discretized Laplacian operator with grid spacing h. Using the exponentiation identity for Kronecker sums, the following Hamiltonian simulation is performed:

$$\begin{aligned} e^{iAt} &= \left(e^{iL_h t} \otimes I \otimes \cdots \otimes I\right)\left(I \otimes e^{iL_h t} \otimes \cdots \otimes I\right) \cdots \left(I \otimes I \otimes \cdots \otimes e^{iL_h t}\right) \\ &= \left(e^{iL_h t} \otimes e^{iL_h t} \otimes \cdots \otimes e^{iL_h t}\right) \end{aligned}$$

Controlled versions of the Hamiltonian simulation are used to kick back the phase, and the remainder of the algorithm proceeds similar to HHL. This algorithm was implemented with modifications for circuit optimization by Wang et al. [13] on a quantum simulator. Childs et al. [7] points out that while the circuit depth scales favorably, the probability of success is $O(\text{poly}(1/\epsilon))$ and finite-difference discretization errors are not considered in the analysis. Childs and Liu [14] approaches the same problem with spectral and adaptive finite difference methods under a "global strict diagonal dominance" requirement, a stricter condition than diagonal dominance, to achieve a complexity of $O\left(d^2 \operatorname{poly}\log(1/\epsilon)\right)$ using the spectral method and $O\left(d^{\frac{13}{2}} \operatorname{poly}\log(d/\epsilon)\right)$ using adaptive finite difference grids. However, the Kronecker product structure does not generalize to unstructured grids on general domains encountered in problems of practical interest.

An algorithm for simulating the wave equation using the finite difference method was presented in [10] for Dirichlet and Neumann boundary conditions using an approach based on a factorization of the Laplacian operator on a rectangular domain as $L_h = BB^\dagger$ to map the problem (second-order time derivative) to the Schrödinger equation (first-order time derivative) solved using Hamiltonian simulation as

$$\frac{\mathrm{d}}{\mathrm{d}t}|\psi(t)\rangle = -\frac{i}{h}\begin{pmatrix} 0 & B \\ B^\dagger & 0 \end{pmatrix}|\psi(t)\rangle$$

by noting that

$$\frac{\mathrm{d}^2}{\mathrm{d}t^2}|\psi(t)\rangle = -\frac{1}{h^2}\begin{pmatrix} BB^\dagger & 0 \\ 0 & B^\dagger B \end{pmatrix}|\psi(t)\rangle = -\frac{1}{h^2}\begin{pmatrix} L_h & 0 \\ 0 & L_h^\dagger \end{pmatrix}|\psi(t)\rangle$$

where $|\psi(t)\rangle = \begin{pmatrix} \phi_V \\ \phi_E \end{pmatrix}$. ϕ_V encodes the solution to the wave equation and ϕ_E are intermediate variables. This algorithm was implemented by Suau et al. [15], albeit using the sub-optimal Trotter–Suzuki method for Hamiltonian simulation.

Engel et al. [11] provide a quantum algorithm for plasma physics problems. A linearized version of the Vlasov equation was derived, which is then evolved in time using Hamiltonian simulation algorithms similar to [10], and a complete circuit description with simulation results is provided with errors scaling as $O(\text{poly}(1/\epsilon))$. Novikau et al. [16] mapped a cold plasma wave model to a Hamiltonian simulation and suggested quantum signal processing for further improvements. These Schrödingerization approaches are specialized and may not generalize to all PDEs.

Algorithms for evolutionary PDEs mentioned above typically scale linearly in simulation time, i.e., $O(t)$. An et al. [12] discretize hyperbolic and parabolic PDEs in space over a rectangular domain to obtain a system of ordinary differential equations. Quantum ODE solvers were then used to evolve the system in time instead of Hamiltonian simulation. They consider two scenarios to "fast-forward" (improve the time-complexity of) the simulation: semidefinite ODE systems with square-root access (similar to the decomposition in [10]) and diagonalizable systems (using a Fourier transform) to get $O(\sqrt{t})$ and $O(\log t)$ scaling in time. The lifting transformation proposed by Costa et al. [10] to transform a second-order ODE system $\frac{\mathrm{d}^2}{\mathrm{d}t^2}u(t) = (A + cI)u(t) + b(t)$ to a first-order ODE system is extended for non-homogeneous systems as

$$\frac{\mathrm{d}}{\mathrm{d}t}\begin{pmatrix} u(t) \\ \tilde{v}(t) \end{pmatrix} = \begin{pmatrix} 0 & I \\ (A + cI) & 0 \end{pmatrix}\begin{pmatrix} u(t) \\ \tilde{v}(t) \end{pmatrix} + \begin{pmatrix} 0 \\ b(t) \end{pmatrix}$$

Analytical results for the transport equation, heat equation, advection–diffusion equation, wave equation, Klein–Gordon equation, Airy equation, and the Euler beam equation are provided. However, the fast-forwarding results are again restricted to specialized domains and structured grids.

Several algorithms have been proposed for numerical solutions of the Navier–Stokes equations for computational fluid dynamics. Quantum lattice-gas models have been proposed by [17–20], with numerical results presented in [21]. However, these methods are for Type-II quantum computers [22] whose architecture differs from the universal gate-based quantum computer architecture. Other approaches, such as the lattice Boltzmann method, have also been mapped to quantum circuits. Budinski [23] detailed circuits and simulations for the advection–diffusion equation, achieving $O(\log_2(\alpha D))$ scaling, where α is the total number of distribution functions and D is the number of distribution functions for each site, albeit with a fixed choice of the relaxation time $\omega = 1$ and a limited choice of D1Q2 and D2Q5 models in 1D and 2D respectively. The approach requires classical computation for renormalization of the post-selected state after each time step. We note, however, that [24] pointed out that the streaming and collision operations are amenable to quantum computation and provide a quantum mapping of transport equations in fluid flows using analogies between the Dirac and Lattice Boltzmann equations to define an algorithm that measures an ancillary qubit at each time step, with a non-zero probability of success.

A quantum Navier-Stokes algorithm based on the quantum amplitude estimation algorithm was proposed by Gaitan [25], which was later generalized to a quantum nonlinear PDE algorithm in [26]. However, the precision of the algorithms scales as $\Omega\left(\frac{1}{\epsilon^{q+1}}\right)$, where $q = r + \rho$ is the smoothness parameter of the solution, with r being the highest derivative retained in a Taylor series expansion and $0 \leq \rho \leq 1$, leading to a $O(\text{poly}(1/\epsilon))$ scaling. Oz et al. [27] improved the accuracy of the algorithm by an order of magnitude by using Chebyshev points, but retained the overall complexity to $O(\text{poly}(1/\epsilon))$. Numerical results were demonstrated using quantum simulators.

References

1. B.D. Clader, B.C. Jacobs, C.R. Sprouse, Preconditioned quantum linear system algorithm. Phys. Rev. Lett. **110**(25), 250504 (2013). https://doi.org/10.1103/PhysRevLett.110.250504
2. A. Montanaro, S. Pallister, Quantum algorithms and the finite element method. Phys. Rev. A **93**(3), 032324 (2016). https://doi.org/10.1103/PhysRevA.93.032324
3. F. Fillion-Gourdeau, E. Lorin, Simple digital quantum algorithm for symmetric first-order linear hyperbolic systems. Numer. Algor. **82**(3), 1009–1045 (2019). https://doi.org/10.1007/s11075-018-0639-3
4. D. Fang, L. Lin, Y. Tong, Time-marching based quantum solvers for time-dependent linear differential equations. Quantum **7**, 955 (2023). https://doi.org/10.22331/q-2023-03-20-955
5. B. Reggio, N. Butt, A. Lytle, P. Draper, Fast partitioning of Pauli strings into commuting families for optimal expectation value measurements of dense operators. Phys. Rev. A **110**(2), 022606 (2024). https://doi.org/10.1103/PhysRevA.110.022606
6. M. Kohda, R. Imai, K. Kanno, K. Mitarai, W. Mizukami, Y.O. Nakagawa, Quantum expectation-value estimation by computational basis sampling. Phys. Rev. Res. **4**(3), 033173 (2022). https://doi.org/10.1103/PhysRevResearch.4.033173
7. A.M. Childs, J.-P. Liu, A. Ostrander, High-precision quantum algorithms for partial differential equations. Quantum **5**, 574 (2021). https://doi.org/10.22331/q-2021-11-10-574
8. D.W. Berry, A.M. Childs, A. Ostrander, G. Wang, Quantum algorithm for linear differential equations with exponentially improved dependence on precision. Commun. Math. Phys. **356**(3), 1057–1081 (2017). https://doi.org/10.1007/s00220-017-3002-y
9. Y. Cao, A. Papageorgiou, I. Petras, J. Traub, S. Kais, Quantum algorithm and circuit design solving the Poisson equation. New J. Phys. **15**(1), 013021 (2013). https://doi.org/10.1088/1367-2630/15/1/013021
10. P.C.S. Costa, S. Jordan, A. Ostrander, Quantum algorithm for simulating the wave equation. Phys. Rev. A **99**(1), 012323 (2019). https://doi.org/10.1103/PhysRevA.99.012323
11. A. Engel, G. Smith, S.E. Parker, Quantum algorithm for the Vlasov equation. Phys. Rev. A **100**(6), 062315 (2019). https://doi.org/10.1103/PhysRevA.100.062315

12. D. An, J.-P. Liu, D. Wang, Q. Zhao, A theory of quantum differential equation solvers: limitations and fast-forwarding (2022). https://doi.org/10.48550/ARXIV.2211.05246
13. S. Wang, Z. Wang, W. Li, L. Fan, Z. Wei, Y. Gu, Quantum fast Poisson solver: the algorithm and complete and modular circuit design. Quantum Inf. Process **19**(6), 170 (2020). https://doi.org/10.1007/s11128-020-02669-7
14. A.M. Childs, J.-P. Liu, Quantum spectral methods for differential equations. Commun. Math. Phys. **375**(2), 1427–1457 (2020). https://doi.org/10.1007/s00220-020-03699-z
15. A. Suau, G. Staffelbach, H. Calandra, Practical quantum computing: solving the wave equation using a quantum approach. ACM Trans. Quantum Comput. **2**(1), 1–35 (2021). https://doi.org/10.1145/3430030
16. I. Novikau, E.A. Startsev, I.Y. Dodin, Quantum signal processing for simulating cold plasma waves. Phys. Rev. A **105**(6), 062444 (2022). https://doi.org/10.1103/PhysRevA.105.062444
17. D.W. Berry, A.M. Childs, R. Cleve, R. Kothari, R.D. Somma, Simulating Hamiltonian dynamics with a truncated Taylor series. Phys. Rev. Lett. **114**(9), 090502 (2015). https://doi.org/10.1103/PhysRevLett.114.090502
18. Y. Shen et al., Estimating Eigenenergies from quantum dynamics: a unified noise-resilient measurement-driven approach (2023). https://doi.org/10.48550/arXiv.2306.01858. arXiv:2306.01858
19. Y. Shen et al., Efficient measurement-driven Eigenenergy estimation with classical shadows (2024). https://doi.org/10.48550/arXiv.2409.13691. arXiv:2409.13691
20. Z. Ding, L. Lin, even shorter quantum circuit for phase estimation on early fault-tolerant quantum computers with applications to ground-state energy estimation. PRX Quantum **4**(2), 020331 (2023). https://doi.org/10.1103/PRXQuantum.4.020331
21. M.M. Micci, J. Yepez, Measurement-based quantum lattice gas model of fluid dynamics in $2 + 1$ dimensions. Phys. Rev. E **92**(3), 033302 (2015). https://doi.org/10.1103/PhysRevE.92.033302
22. J. Yepez, Type-II quantum computers. Int. J. Mod. Phys. C **12**(09), 1273–1284 (2001). https://doi.org/10.1142/S0129183101002668
23. L. Budinski, Quantum algorithm for the advection–diffusion equation simulated with the lattice Boltzmann method. Quantum Inf. Process **20**(2), 57 (2021). https://doi.org/10.1007/s11128-021-02996-3
24. A. Mezzacapo, M. Sanz, L. Lamata, I.L. Egusquiza, S. Succi, E. Solano, Quantum simulator for transport phenomena in fluid flows. Sci. Rep. **5**(1), 13153 (2015). https://doi.org/10.1038/srep13153
25. F. Gaitan, Finding flows of a Navier–Stokes fluid through quantum computing. npj Quantum Inf. **6**(1), 61 (2020). https://doi.org/10.1038/s41534-020-00291-0
26. F. Gaitan, Finding solutions of the Navier-Stokes equations through quantum computing—recent progress, a generalization, and next steps forward. Adv. Quantum Tech. **4**(10), 2100055 (2021). https://doi.org/10.1002/qute.202100055
27. F. Oz, O. San, K. Kara, An efficient quantum partial differential equation solver with Chebyshev points. Sci. Rep. **13**(1), 7767 (2023). https://doi.org/10.1038/s41598-023-34966-3

35 Variational Algorithms: Theory

With the exception of the Hamiltonian simulation of sparse systems using low-order Trotter–Suzuki methods and short evolution times for a small number of qubits, or Hadamard tests with significant circuit optimization, none of the algorithms discussed in the previous chapters can run on contemporary NISQ hardware. This is due to high sensitivity to noise, hardware imperfections, and limited physical qubit counts that prevent error correction. In the NISQ era, there is strong demand for algorithms that can provide quantum advantage using noisy hardware. Variational algorithms have recently gained attention as a hybrid classical-quantum approach for algorithms that can operate on NISQ devices. Examples include the Quantum Adiabatic Optimization Algorithm (QAOA) [1] for combinatorial problems, Variational Quantum Linear Solver (VQLS) [2] for linear systems, and the Variational Quantum Eigensolver [3] ground-state preparation in quantum chemistry problems. However, the classical-quantum training loop of variational algorithms is costly and currently precludes quantum advantage. Optimization is problem- and hardware-specific and is not known to generalize. For example, [4] shows that VQLS, when trained to solve a 1D Poisson's problem, does not generalize to finer discretizations.

Variational quantum algorithms are analogous to classical machine learning models, but use a quantum ansatz rather than a classical one, and use quantum circuits to evaluate the cost function.

The core ingredients of a variational quantum algorithm are a cost function $C(\theta)$, a parametrized quantum ansatz $U(\theta)$, and an optimization method (gradient-based or gradient-free) [5] as shown in Fig. 35.1.

The cost function encodes the solution of the problem as

$$\theta^* = \arg\min_{\theta} C(\theta)$$

O. M. Raisuddin and S. De, *Quantum Computing for Engineers*,
https://doi.org/10.1007/978-3-032-03325-3_35

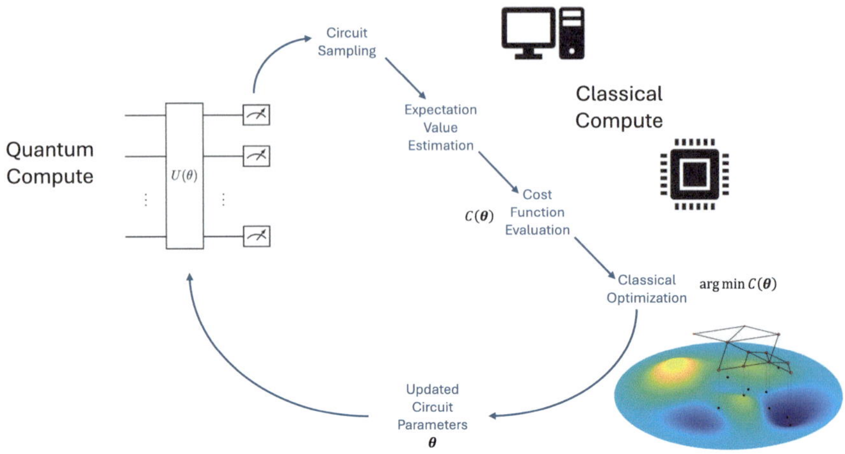

Fig. 35.1 Schematic of a variational quantum algorithm optimization loop

where θ is a set of trainable parameters of an ansatz $U(\theta)$, a trainable quantum circuit parametrized by θ. The parameters θ can be continuous or discrete. As an example, a continuous parameter could be the rotation angle for a Y rotation gate, and a discrete parameter could be whether to apply a quantum gate in a circuit or not. The cost function can be expressed as

$$C(\theta) = f\{\{\rho_k\}, \{O_k\}, U(\theta)\}$$

where f is some function, $\{\rho_k\}$ are input states (training data), and $\{O_k\}$ are observables, or measurements of the quantum circuit. The evaluation of $C(\theta)$ is performed using quantum computers (possibly with some classical post-processing), while the optimization of θ is performed using classical computers, leading to a hybrid algorithm. For NISQ hardware, the quantum circuit must fit hardware constraints, i.e., limited qubits and shallow depth. Some approaches work directly with hardware-level optimization, e.g., microwave pulse shaping for superconducting qubits.

The choice of the ansatz determines the parameters θ. In general, a quantum ansatz can be expressed as

$$U(\theta) = \prod_{l=1}^{L} U_l(\theta_l)$$

Many different ansatze have been proposed for various problems, with the common goal of efficiency and trainability on NISQ hardware.

The cost function may be optimized using gradient-based or gradient-free approaches. Gradient-based approaches compute the derivatives $\frac{\partial C(\theta_l)}{\partial \theta_l}$. The derivative may be approximated using finite differencing. However, "parameter shift" rules [6–9] allow exact differentiation of parametrized quantum circuits (for continuous parameters) analogous to automatic differentiation for classical computing. The key idea behind parameter shift rules is that to compute $\frac{\partial C(\theta_l)}{\partial \theta_l}$, the same quantum circuit is used with a "parameter shift" $s \in \mathbb{R}$ applied to the parameter θ_l with a multiplier $c \in \mathbb{R}$:

$$\frac{\partial C(\theta)}{\partial \theta_l} = c\left[C\left(\theta^{(l)}\right) - C\left(\theta^{(l)}\right)\right]$$

where

$$\theta_k^{(l)} = \begin{cases} \theta_k, & \forall\, k \neq l \\ \theta_k + s & k = l \end{cases}$$

Although similar to finite differences in appearance, the parameter shift rule is exact. Higher order derivatives may be computed by nesting this rule. Parameter shift rules can be computed for any arbitrary circuit with continuous parameters. Hybrid quantum-classical models can combine parameter shift for quantum circuits with automatic differentiation for classical parts [5, 10].

However, optimization of quantum circuits can suffer from vanishing gradients [10, 11]. In these cases, classical derivative-free optimizers like Nelder–Mead [12, 13] can be used if the parameter space is not too large. Optimization is also sensitive to noise; recent studies suggest SPSA and CMA-ES are more robust in noisy settings [14].

References

1. E. Farhi, J. Goldstone, S. Gutmann, A quantum approximate optimization algorithm. (2014). arXiv. https://doi.org/10.48550/ARXIV.1411.4028
2. C. Bravo-Prieto, R. LaRose, M. Cerezo, Y. Subasi, L. Cincio, and P. J. Coles, Variational quantum linear solver. (2019). arXiv. https://doi.org/10.48550/ARXIV.1909.05820
3. P. J. Ollitrault et al., Quantum equation of motion for computing molecular excitation energies on a noisy quantum processor. Phys. Rev. Research, **2**(4), 043140(2020). https://doi.org/10.1103/PhysRevResearch.2.043140
4. E. Cappanera, Variational quantum linear solver for finite element problems: a Poisson equation test case. TU Delft, 2021. [Online]. http://resolver.tudelft.nl/uuid:deba389d-f30f-406c-ad7b-babb1b298d87
5. M. Cerezo et al., Variational quantum algorithms. Nat Rev Phys., **3**(9), 625–644 (2021). https://doi.org/10.1038/s42254-021-00348-9
6. N. Linden, A. Montanaro, C. Shao, Quantum vs. classical algorithms for solving the heat equation. Commun. Math. Phys., **395**(2), 601–641 (2022). https://doi.org/10.1007/s00220-022-04442-6
7. I. Kerenidis, A. Prakash, Quantum gradient descent for linear systems and least squares. Phys. Rev. A, **101**(2), 022316 (2020). https://doi.org/10.1103/PhysRevA.101.022316
8. P. Rebentrost, M. Schuld, L. Wossnig, F. Petruccione, S. Lloyd, Quantum gradient descent and Newton's method for constrained polynomial optimization. New J. Phys., **21**(7), 073023 (2019). https://doi.org/10.1088/1367-2630/ab2a9e

9. O. M. Raisuddin, S. De, Quantum multigrid algorithm for finite element problems. (2024). https://doi.org/10.48550/ARXIV.2404.07466
10. M. Broughton et al., TensorFlow quantum: A software framework for quantum machine learning. (2020). arXiv. https://doi.org/10.48550/ARXIV.2003.02989
11. S. Khatri, R. LaRose, A. Poremba, L. Cincio, A. T. Sornborger, P. J. Coles, Quantum-assisted quantum compiling. Quantum, **3**, 140(2019). https://doi.org/10.22331/q-2019-05-13-140
12. J. R. McClean, J. Romero, R. Babbush, A. Aspuru-Guzik, The theory of variational hybrid quantum-classical algorithms. New J. Phys., **18**(2), 023023(2016). https://doi.org/10.1088/1367-2630/18/2/023023
13. J. A. Nelder, R. Mead, A simplex method for function minimization. Comp. Journal., **7**(4), 308–313(1965). https://doi.org/10.1093/comjnl/7.4.308
14. X. Bonet-Monroig et al., Performance comparison of optimization methods on variational quantum algorithms. Phys. Rev. A, **107**(3), 032407 (2023). https://doi.org/10.1103/PhysRevA.107.032407

Notable Variational Algorithms: VQE, QAOA, and VQLS

36

In this chapter, we provide an overview of variational quantum algorithms for eigenvalue, combinatorial, and linear system problems.

Variational Quantum Eigensolver

The most notable variational quantum algorithm is the variational quantum eigensolver (VQE), which was introduced by [1]. VQE seeks the extremal eigenvalues of a Hamiltonian by minimizing a cost function defined as an expectation value. This approach is fundamental for preparing ground states of quantum systems and estimating their properties. As an example, the VQE has been used to compute and prepare the ground and excited states and energies of molecules [2–5], which is essential in quantum chemistry.

The VQE cost function is defined as

$$\langle\psi(\theta)|H|\psi(\theta)\rangle = \langle 0|U(\theta)^{\dagger}HU(\theta)|0\rangle$$

where H is typically a sum of Pauli strings, and $U(\theta)$ is a parametrized quantum circuit. In quantum chemistry, $U(\theta)$ is often chosen to be a unitary coupled cluster (UCC) circuit (or one of its variants). For NISQ devices and general problems, hardware-efficient ansatz such as SU(n) are commonly chosen.

Below, we provide an example where VQE is used to solve a random Hamiltonian expressed as a sum of Pauli strings, employing a hardware-efficient SU(2) ansatz. Convergence results are shown in Fig. 36.1:

```
#!/usr/bin/python3
```


O. M. Raisuddin and S. De, *Quantum Computing for Engineers*,
https://doi.org/10.1007/978-3-032-03325-3_36

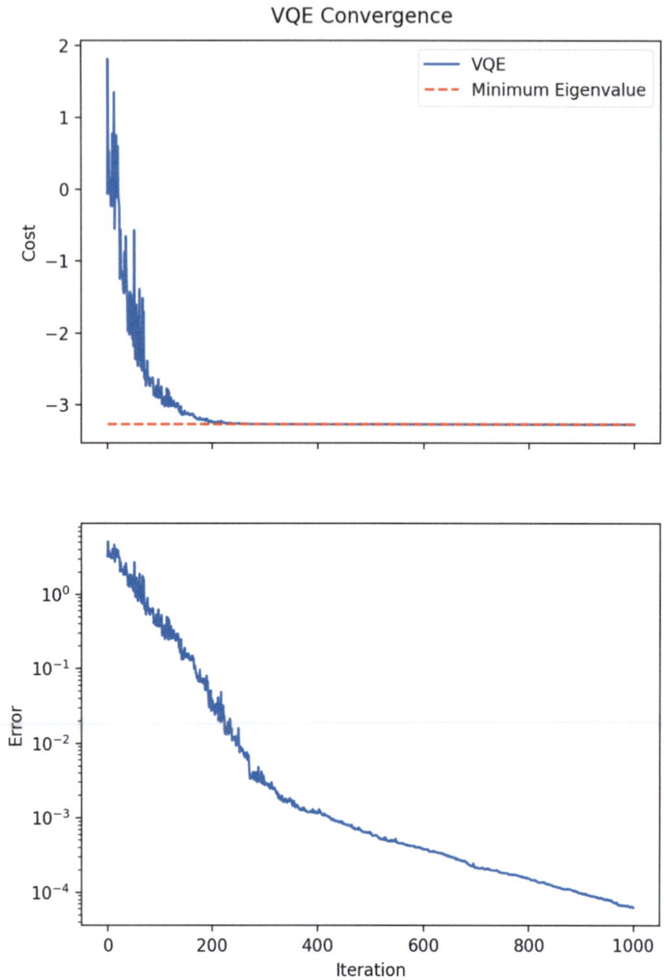

Fig. 36.1 Convergence of the variational quantum eigensolver

```
import numpy as np
import matplotlib.pyplot as plt
from qiskit.circuit.library import EfficientSU2
from qiskit.quantum_info import random_pauli, SparsePauliOp
from qiskit.primitives import StatevectorEstimator
from scipy.optimize import minimize

# Define number of qubits
n = 3
```

```
# Define Hamiltonian as a sum of Pauli strings
num_paulis = 10
mypaulis = [SparsePauliOp(random_pauli(n), np.random.rand()) for
_i in range(num_paulis)]
hamiltonian = sum(mypaulis)

# Create VQE ansatz
ansatz = EfficientSU2(num_qubits=n,reps=3)

# Define cost function
def cost_function_generator(circuit,observables):
    def cost_function(params):
        estimator = StatevectorEstimator()
        pub = (circuit, observables, params)
        job = estimator.run([pub])
        cost = job.result()[0].data['evs']
        return cost
    return cost_function

mycostfunction = cost_function_generator(ansatz,hamiltonian)

# Define a callback function to track progress of optimization
cost_history = []
def callback_function_generator(cost_history, cost_function):
    def callback(theta):
        cost = cost_function(theta)
        cost_history.append(cost)
        return None
    return callback

mycallback = callback_function_generator(cost_history, mycost-
function)

# Create initial guess
theta = np.zeros(ansatz.num_parameters)

# Optimize ansatz
result = minimize(mycostfunction,theta,method='COBYLA',
callback=mycallback)

# Get exact minimum eignevalue
exact_min_eig = min(np.linalg.eig(hamiltonian.to_
matrix())[0]).real
```

```
# Compute errors
error = cost_history - exact_min_eig

# Plot results
fig, axs = plt.subplots(2,sharex=True)
fig.suptitle('VQE Convergence')
axs[0].plot(cost_history)
axs[0].hlines(exact_min_eig,xmin=0,xmax=len(cost_
history),colors='r',linestyles='dashed')
axs[0].legend(['VQE', 'Minimum Eigenvalue'])
axs[0].set(ylabel='Cost')
axs[1].plot(error)
axs[1].set_yscale('log')
axs[1].set(xlabel='Iteration',ylabel='Error')
plt.show()
```

Variational Quantum Linear Solver

Variational quantum linear solvers (VQLS) [6] have recently gained significant attention. To solve a linear system $Ax = b$ using the VQLS, the system matrix needs to be provided as a sum of unitaries:

$$A = \sum_k \alpha_k U_k$$

and the vector b should be accessible via a unitary state preparation.

$$U_b|0\rangle = |b\rangle.$$

A parametrized quantum circuit $U(\theta)$ is chosen, with the goal of optimizing θ such that

$$U\left(\theta_{\text{opt}}\right)|0\rangle = |x(\theta)\rangle \approx |x\rangle$$

The global cost function for the VQLS is defined as

$$\tilde{C}_G = \langle x(\theta)A^\dagger|(I - |b\rangle\langle b|)A|x(\theta)\rangle = \langle x(\theta)|A^\dagger A|x(\theta)\rangle - \langle x(\theta)|A^\dagger|b\rangle\langle b|A|x(\theta)\rangle$$

When $\langle x(\theta)|A^\dagger A|x(\theta)\rangle$ is small, C is also small, even if $|x(\theta)\rangle$ is not close to $|b\rangle$. To avoid this, the cost is normalized by $\langle x(\theta)|A^\dagger A|x(\theta)\rangle$ to get

$$C_G = \frac{\langle x(\theta)|A^\dagger A|x(\theta)\rangle}{\langle x(\theta)|A^\dagger A|x(\theta)\rangle} - \frac{\langle x(\theta)|A^\dagger|b\rangle\langle b|A|x(\theta)\rangle}{\langle x(\theta)|A^\dagger A|x(\theta)\rangle} = 1 - \frac{\langle|\langle x(\theta)|A^\dagger|b\rangle|^2}{\langle x(\theta)|A^\dagger A|x(\theta)\rangle}$$

This formulation requires evaluating $\langle x(\theta)|A^\dagger|b\rangle$ and $\langle x(\theta)|A^\dagger A|x(\theta)\rangle$ which can be computed using the SWAP (or Hadamard) test:

$$\langle x(\theta)|A^\dagger A|x(\theta)\rangle = \sum_{mn} \alpha_m^* \alpha_n \langle x(\theta)|U_m^\dagger U_n|x(\theta)\rangle$$

$$\left|\langle x(\theta)|A^\dagger|b\rangle\right|^2 = \sum_{mn} \alpha_m^* \alpha_n \langle x(\theta)|U_m^\dagger|b\rangle\langle b|U_n|x(\theta)\rangle$$

To address convergence issues related to barren plateaus, the global cost function is further modified by introducing multiple local cost functions that indirectly minimize the global cost function [6]:

$$C_L = \langle x(\theta)|\left(A^\dagger U_b\left(I - \frac{1}{n}\sum_{j=1}^{n}|0_j\rangle\langle 0_j| \otimes I_{\bar{j}}\right)U^\dagger A\right)|x(\theta)\rangle$$

where $|0_j\rangle$ is the $|0\rangle$ state on qubit j and $I_{\bar{j}}$ is the identity on all qubits except qubit j.

These cost functions are bounded as

$$C_G, nC_L \geq \frac{\epsilon^2}{\kappa^2}$$

VQLS has been applied to solve the heat equation in 1D [7, 8], in 2D [9], and to potential and Stokes flow in 2D [10]. [9, 10] demonstrate logarithmic scaling in $1/\in$ and N. However, they use the Pauli basis for their matrix, which can have $O(N)$ terms for a 1D discrete Laplacian. [11] provides an efficient tensor product decomposition with $O(\log(N))$ terms.

Quantum Approximate Optimization Algorithm

The quantum approximate optimization algorithm (QAOA) [12] is a variational algorithm designed for combinatorial optimization problems. The problem is typically formulated as a Quadratic Unconstrained Binary Optimization (QUBO) problem:

$$\arg\min_{x\in\{0,1\}^n} x^T Q x + x^T b$$

$x \in \{0, 1\}^n$ is a binary string of n bits, $Q \in \mathbb{R}^{N\times N}$ is a matrix, and $b \in \mathbb{R}^N$ encoding a combinatorial problem. Note that b_i can be absorbed into Q_{ii}.

By substituting $x_i = \frac{1-Z_i}{2}$ where Z_i is a Pauli-Z operator on the i-th qubit, the problem can be reformulated as the Ising Hamiltonian problem:

$$H_{Ising} = \sum_{i,j} Z_i Z_j J_{ij} + \sum_i Z_i h_i$$

$$\underset{|\psi\rangle\in\{-1,1\}^{\otimes n}}{\arg\min}\ \langle\psi|H_{\text{Ising}}|\psi\rangle$$

where $|\psi \in \{-1, 1\}^{\otimes n}$ is a string of spin-up (+1) and spin-down (–1) states in the $\{|+z\rangle, |-z\rangle\}$ basis, $J \in \mathbb{R}^{N\times N}$ is a matrix, and $h \in \mathbb{R}^N$ is a vector.

Both QUBO and Ising Hamiltonian optimization are NP-hard. Classical heuristics are used for such problems, and QAOA is also a heuristic method, inspired by the quantum adiabatic theorem and its discretization.

For clarity, we will illustrate QAOA using the Max-Cut problem as an example. First, we provide an overview of the quantum adiabatic theorem, then introduce the Max-Cut problem, and finally formulate QAOA for Max-Cut.

Quantum Adiabatic Theorem

The quantum adiabatic theorem states that if a quantum system in its ground state evolves slowly enough, it will remain in its ground state. This can be formulated mathematically by considering an initial Hamiltonian H_I with a known ground state, and a final Hamiltonian H_F with an unknown ground state. The evolution of such a system can be modeled using a time-dependent Hamiltonian $H(s)$.

$$H(s) = A(s)H_I + B(s)H_F$$

where $s \in [0, 1]$ quantifies the transition of the Hamiltonian from H_I to H_F by using "scheduling functions" $A(s)$ and $B(s)$ s.t. $A(0) \gg B(0)$ and $A(1) \ll B(1)$. An example of such scheduling functions is given in Fig. 36.2.

Denoting the known ground state of H_I as $|\psi(0)\rangle$, according to the Schrodinger equation this state will evolve as

$$H(s)|\psi(s)\rangle = i\frac{\partial}{\partial t}|\psi(s)\rangle$$

Evolving this system adiabatically, or infinitesimally slowly by assigning $s = 0 : t = 0$ and $s = 1 : \lim t \to \infty$, will guarantee that $|\psi(1)\rangle$ will be a ground state

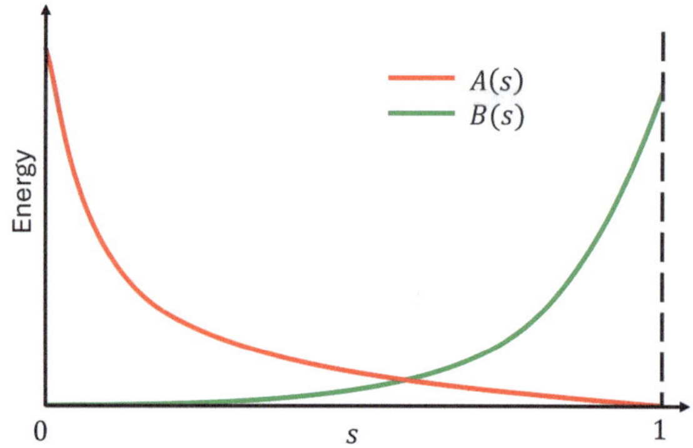

Fig. 36.2 Example of a schedule function

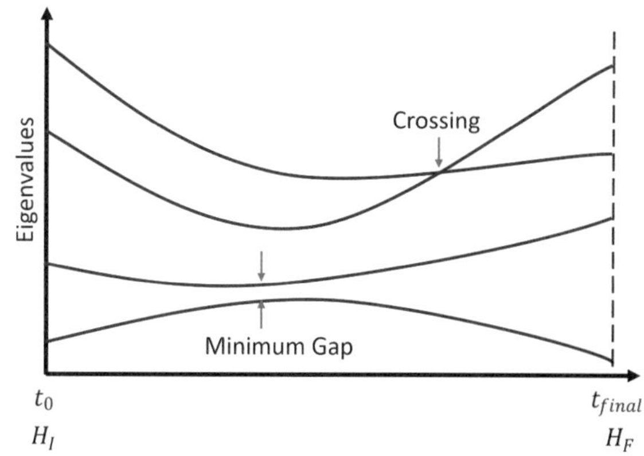

Fig. 36.3 Evolution of the eigenvalues for $H(s)$ with one crossing and the minimum gap labeled

of H_F. Figure 36.3 presents a visualization of the transition of eigenstates from H_I to H_F.

Such infinitesimal transitions are not practical. They are dependent on the minimum gap between the ground state and first excited state, as shown in Fig. 36.3, which may not be known either. Instead, one may develop a heuristic technique by attempting to *approximately* solve this problem by evolving the quantum state in finite time. This will result in a loss of the guaranteed ground state, but it may still yield high-quality solutions. Furthermore, this evolution can be approximated on a digital quantum computer using Trotterization, which will introduce some discretization errors. However, in the limit of infinitely many Trotter steps, the guarantee can be recovered.

We finally note that the initial Hamiltonian and final Hamiltonian are also referred to as the mixer Hamiltonian and the cost Hamiltonian, respectively.

Weighted Max-Cut Problem

The Weighted Max-Cut problem is roughly described as a partitioning of a graph into two disjoint graphs. The Max-Cut problem is a specific instance with equal weights for all edges.

We define a graph G as $G = \{V, E, W\}$ as a set of vertices V, edges E connecting V, and a set of weights W associated with each edge. The Max-Cut problem is to form a bipartite partition of V, i.e., V_A and V_B s.t. $V_A \cup V_B = V$, $V_A \cap V_B = \emptyset$, such that the sum of the weights of edges between V_A and V_B is maximized.

As an example, consider the graph shown in Fig. 36.4 with 6 vertices and 9 edges with associated weights.

The Max-Cut solution for this problem is shown in Fig. 36.5, with the edges contributing to the sum highlighted in red. Finding the Max-Cut solution for a graph is NP-hard. However, classical heuristic approaches can find approximate or good-quality solutions.

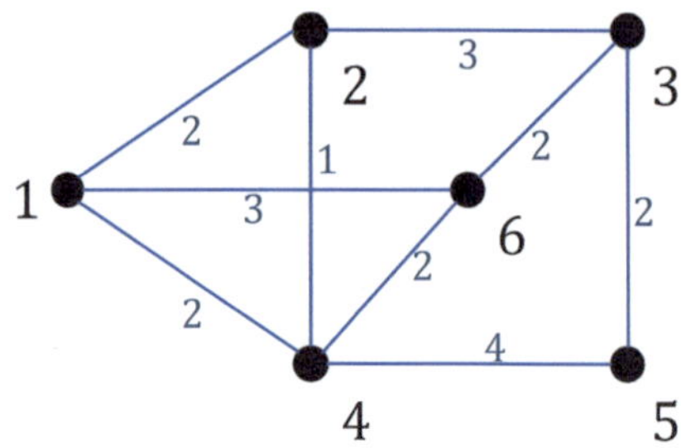

Fig. 36.4 Problem graph for a weighted Max-Cut problem

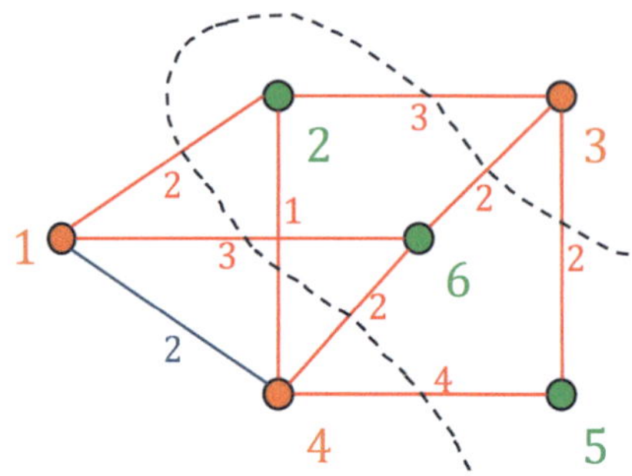

Fig. 36.5 Optimal cut with the two sets of vertices colored

We can formulate this problem mathematically by assigning 0, 1 to the vertices in V_A, V_B respectively, which leads to the cost function

$$C(x) = \sum_{i,j=0}^{n-1} W_{ij} x_i \left(1 - x_j\right) = -\sum_{i,j=0}^{n-1} W_{ij} x_i x_j + \sum_{i,j=0}^{n-1} W_{ij} x_i$$

where W is a matrix with entries W_{ij} containing the weights associated with the edge connecting vertices i and j.

For the problem shown in Fig. 36.5, the weight matrix and two equivalent optimal solutions (with $C\left(x_{opt}\right) = 19$) can be written as

$$W = \begin{pmatrix} 0 & 2 & & 4 & & 3 \\ 2 & 0 & 3 & 1 & & \\ & 3 & 0 & & 2 & 2 \\ 4 & 1 & & 0 & 4 & 2 \\ & & 2 & 4 & 0 & \\ 3 & & 2 & 2 & & 0 \end{pmatrix}, \; x = \begin{pmatrix} 0 \\ 1 \\ 0 \\ 0 \\ 1 \\ 1 \end{pmatrix}, \begin{pmatrix} 1 \\ 0 \\ 1 \\ 1 \\ 0 \\ 0 \end{pmatrix}$$

This is readily translated to a QUBO problem with $Q_{ij} = -W_{ij}$ and $b_i = \sum_{i,j=0}^{n-1} W_{ij}$. In the QAOA, we want the cost function to take the form

$$\langle \psi | H_C | \psi \rangle = c(\psi)$$

where H_C is a cost Hamiltonian that is acting on $|\psi\rangle$. To apply the quantum adiabatic theorem, we transform the QUBO problem into an Ising Hamiltonian such that

$$H_C|\psi\rangle = c(\psi)|\psi\rangle$$

where H_C is the cost (or final) Hamiltonian and $c(\psi)$ is the cost associated with the state $|\psi\rangle$. To achieve this, we substitute $x_i = \frac{1-Z_i}{2}$, converting the QUBO variables into quantum operators acting on $|\psi\rangle$.

By substituting $x_i = \frac{1-Z_i}{2}$ in $C(x)$, we get

$$C(x) = \frac{1}{4}\sum_{i,j=0}^{n-1} Q_{ij}Z_iZ_j - \frac{1}{2}\sum_{i=1}^{n-1}\left(b_i + \sum_{j=1}^{n-1} Q_{ij}\right)Z_i + \left(\frac{1}{4}\sum_{i,j=0}^{n-1} Q_{ij} + \frac{1}{2}\sum_{i=0}^{n-1} b_i\right)$$

Since the constant terms $\left(\frac{1}{4}\sum_{i,j=0}^{n-1} Q_{ij} + \frac{1}{2}\sum_{i=0}^{n-1} b_i\right)$ do not contribute to the optimization problem, we may simply drop them to arrive at the cost Hamiltonian

$$H_C = \frac{1}{4}\sum_{i,j=0}^{n-1} Q_{ij}Z_iZ_j - \frac{1}{2}\sum_{i=1}^{n-1}\left(b_i + \sum_{j=1}^{n-1} Q_{ij}\right)Z_i$$

We want to evolve a quantum state according to this Hamiltonian, as we have discussed previously while introducing the quantum adiabatic theorem. The Hamiltonian H_C is in a form that can be readily Trotterized as

$$e^{-i\gamma H_C} \approx \left(\prod_{i,j=0}^{n-1} e^{\frac{-i\gamma}{4}Q_{ij}Z_iZ_j}\right)\left(\prod_{i=1}^{n-1} e^{\frac{i\gamma}{2}\left(b_i+\sum_{j=1}^{n-1} Q_{ij}\right)Z_i}\right) = U_{ZZ}(\gamma)U_Z(\gamma) = U_{H_C}(\gamma)$$

We note that exponential terms of the form $e^{-i\alpha Z_iZ_j}$ can be implemented using controlled R_Z gates as discussed in Chap. 28: Hamiltonian Simulation Techniques.

QAOA

We now have most of the ingredients for the QAOA. We provide a recap before proceeding with the remainder. The objective is to optimize a cost function corresponding to a combinatorial problem, in this case a Max-Cut problem. We approximately optimize this cost function using a finite-time version of the quantum adiabatic theorem, which requires an initial (mixer) Hamiltonian and a final (cost) Hamiltonian. The process begins with the known ground state of the initial Hamiltonian and evolves it toward the unknown ground state (optimal solution) of the final Hamiltonian. The cost Hamiltonian has been discussed earlier; the missing piece is the initial Hamiltonian (and its ground state) and the algorithm.

A commonly used mixer Hamiltonian is

$$H_M = \sum_{i=0}^{n-1} X_i$$

where the I_j gate is implied for $i \neq j$.

This Hamiltonian can be exactly exponentiated (since the terms commute) as

$$e^{-i\beta H_M} = \prod_{i=1}^{n-1} e^{-i\beta X_i}$$

It can also be diagonalized using Hadamard gates:

$$H_M = \sum_{i=0}^{n-1} X_i = \sum_{i=0}^{n-1} H_i Z_i H_i = H^{\otimes n} \left(\sum_{i=0}^{n-1} Z_i \right) H^{\otimes n}$$

where H is the Hadamard gate and Z_i is the Pauli-Z operator. The ground state is the uniform superposition:

$$H^{\otimes n} |0\rangle^{\otimes n}$$

Given the parametrized mixer Hamiltonian and cost Hamiltonian, we can Trotterize the time evolution in discrete steps. The QAOA algorithm proceeds as follows:

1. Prepare the ground state $H^{\otimes n}|0$ of the mixer Hamiltonian.
2. Apply the mixer and cost Hamiltonians in alternating fashion for l layers, each parameterized by $\gamma_1, \gamma_2, \ldots, \gamma_l$ and $\beta_1, \beta_2, \ldots, \beta_l$.
3. Update parameters to optimize the expected cost function value.
4. Measure the qubits at the end to obtain candidate solutions as bitstrings.

In the limit of infinite layers or infinite Trotter steps, the adiabatic theorem can be recovered, which guarantees the optimal solution. There is a rich literature on the QAOA and techniques for initializing parameters, which are beyond the scope of this book.

We provide here an example of the QAOA algorithm applied to the weighted Max-Cut problem described earlier.

Using the weight matrix $W = -Q$ provided above, we can form the cost Hamiltonian:

$$H_C = \frac{1}{4} \sum_{i,j=0}^{n-1} Q_{ij} Z_i Z_j - \frac{1}{2} \sum_{i=1}^{n-1} \left(b_i + \sum_{j=1}^{n-1} Q_{ij} \right) Z_i$$

The sampled bitstrings with their quasi-probabilities and cost are plotted in Fig. 36.6:

```
#!/usr/bin/python3

from qiskit.quantum_info import SparsePauliOp
from qiskit.circuit.library import QAOAAnsatz
from qiskit.circuit import QuantumCircuit
```

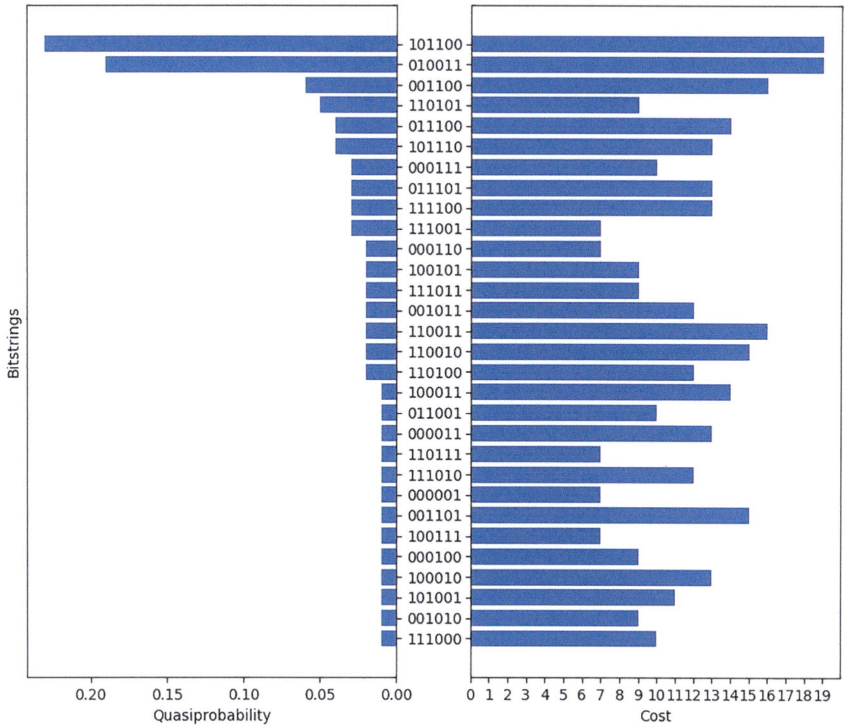

Fig. 36.6 Results for the QAOA applied to a Max-Cut problem. The optimal cut has a high probability of being sampled

```
import numpy as np
from scipy.optimize import minimize
from qiskit.primitives import StatevectorEstimator as Estimator
from qiskit.primitives import StatevectorSampler as Sampler
from matplotlib import pyplot as plt

# Create a dictionary of all the weights for the vertices
weights = {(1,2):2, (1,4):2, (1,6):3, (2,4):1, (2,3):3, (3,6):2,
(3,5):2, (4,6):2, (4,5):4}

# Create weight matrix W
n = max([max(key) for key in weights.keys()])
W = np.zeros((n,n))
for key in weights:
      i,j = key
      W[i-1,j-1] = weights[key]
      W[j-1,i-1] = weights[key]
```

```
# Create QUBO problem from weight matrix
Q = -W
b = np.zeros((n,1))
for i in range(n):
    b[i] = np.sum(W[i,:])

# Form Cost Hamiltonian from QUBO problem
# This will be an Ising Hamiltonian
string_list = []
coeff_list = []
for i in range(n):
    for j in range(n):
        if Q[i,j]!=0:
            string = 'I'*n
            string = string[:i] + 'Z' + string[i+1:]
            string = string[:j] + 'Z' + string[j+1:]
            string_list.append(string)
            coeff_list.append(Q[i,j]/4)
for i in range(n):
    coeff = -b[i]/2
    for j in range(n):
        coeff += -Q[i,j]/2
    string = 'I'*n
    string = string[:i] + 'Z' + string[i+1:]
    if coeff!=0:
        string_list.append(string)
        coeff_list.append(coeff)
H_c = SparsePauliOp(string_list,coeff_list)
# At this point using the following:
# circuit = QAOAAnsatz(cost_operator=H_c, reps=2)
# will be sufficient. The following lines prepare
# the mixer Hamiltonian and initial state for
# completeness of demonstration.

# Form mixer Hamiltonian
string_list = []
coeff_list = []
for i in range(n):
    string = 'I'*n
    string = string[:i] + 'X' + string[i+1:]
    string_list.append(string)
    coeff_list.append(1)
H_m = SparsePauliOp(string_list,coeff_list)

# Quantum circuit to prepare initial state
```

```
initial_state = QuantumCircuit(n)
initial_state.h(range(n))

# Create QAOA ansatz
circuit = QAOAAnsatz(cost_operator=H_c, mixer_operator=H_m, ini-
tial_state=initial_state, reps=5)

# Define QAOA cost function,
# return with -ve sign since we are maximizing using a scipy minimizer
def QAOA_cost(parameters, circuit, H_c, estimator):
    pub = (circuit, H_c, parameters)
    job = estimator.run([pub])
    result = job.result()[0]
    cost = result.data.evs
    cost_history.append(cost)
    return -cost

# Set up Estimator primitive for optimization
estimator = Estimator()
# Track optimization progress
cost_history = []
# Guess initial parameters
init_params = np.ones(len(circuit.parameters))

# Maximize the cost (cost returns -ve)
result = minimize(QAOA_cost,
                  init_params,
                  args=(circuit, H_c, estimator),
                  method='L-BFGS-B',
                  tol=1e-5,
                 )

# Set up sampler primitive to get optimized results
sampler = Sampler()

# Set up optimized circuit for sampling
optimized_circuit = circuit.assign_parameters(result.x)
optimized_circuit.measure_all()
pub = (optimized_circuit)
# Sample circuit
shots = 100
job = sampler.run([pub],shots=shots)
counts = job.result()[0].data.meas.get_counts()
```

```
# Sort by number of counts
sorted_counts = [(key, counts[key]) for key in sorted
(counts,key=counts.get)]

# Compute costs of each counts bitstring
def get_cost_from_string(string, Q, b):
    x = [int(i) for i in string]
    cost = np.einsum('i,ij,j->',x,Q,x) + np.einsum('i,ij->',x,b)
    return cost
costs = [(key,get_cost_from_string(key,Q,b)) for (key,value) in
sorted_counts]

# Plot results
fig, axes = plt.subplots(1,2, sharey=True, figsize=(10, 8))
axes[0].barh([i for (i,j) in sorted_counts],[j/shots for (i,j) in
sorted_counts], align='center')
axes[0].invert_xaxis()
axes[0].set_xlabel('Quasiprobability')
axes[0].set_ylabel('Bitstrings')
axes[0].yaxis.tick_right()
axes[1].barh([i for (i,j) in costs],[j for (i,j) in costs],
align='center')
axes[1].set_xlabel('Cost')
axes[1].set_xticks(list(range(0,20,1)))
plt.show()
```

Variational algorithms are flexible and can be applied to a variety of other problems. For example, [13] presents a variational fast-forwarding technique that uses a Trotter circuit to train a fast-forwardable variational ansatz, reducing simulation time and enabling longer Hamiltonian simulations on NISQ hardware. Variational algorithms have also been explored for financial applications [14], cosmological simulations [15], vehicle routing problems [16], and nonlinear PDEs [17, 18]. For instance, [4] demonstrates the use of multiple copies of variational quantum states to treat nonlinearities, showing exponential efficiency over matrix product states and presenting experimental results. [5] introduces a Chebyshev feature map for nonlinear PDEs, with simulation results for Navier–Stokes equations.

However, the classical cost of optimizing variational circuit parameters can limit quantum advantage for some problems. In the worst case, optimizing variational quantum ansatze is NP-hard in general due to exponentially increasing local minima with the number of parameters and the optimization landscape exhibiting barren plateaus and narrow gorges [19]. Despite these drawbacks, variational quantum algorithms are still expected to find utility in the fault-tolerant regime, e.g., as a method for preparing approximate ground states for quantum phase estimation.

References

1. A. Peruzzo *et al.*, "A variational eigenvalue solver on a photonic quantum processor," *Nat Commun*, vol. 5, no. 1, p. 4213, Jul. 2014, https://doi.org/10.1038/ncomms5213
2. P. J. Ollitrault *et al.*, "Quantum equation of motion for computing molecular excitation energies on a noisy quantum processor," *Phys. Rev. Research*, vol. 2, no. 4, p. 043140, Oct. 2020, https://doi.org/10.1103/PhysRevResearch.2.043140
3. S. Gocho *et al.*, "Excited state calculations using variational quantum eigensolver with spin-restricted ansätze and automatically-adjusted constraints," *npj Comput Mater*, vol. 9, no. 1, p. 13, Jan. 2023, https://doi.org/10.1038/s41524-023-00965-1.
4. A. Kandala *et al.*, "Hardware-efficient variational quantum eigensolver for small molecules and quantum magnets," *Nature*, vol. 549, no. 7671, pp. 242–246, Sep. 2017, https://doi.org/10.1038/nature23879
5. W. J. Huggins, J. Lee, U. Baek, B. O'Gorman, and K. B. Whaley, "A non-orthogonal variational quantum eigensolver," *New J. Phys.*, vol. 22, no. 7, p. 073009, Jul. 2020, https://doi.org/10.1088/1367-2630/ab867b
6. C. Bravo-Prieto, R. LaRose, M. Cerezo, Y. Subasi, L. Cincio, and P. J. Coles, "Variational quantum linear solver," 2019, *arXiv*. https://doi.org/10.48550/ARXIV.1909.05820.
7. E. Cappanera, "Variational quantum linear solver for finite element problems: a Poisson equation test case," TU Delft, 2021. [Online]. Available: http://resolver.tudelft.nl/uuid:deba389d-f30f-406c-ad7b-babb1b298d87
8. C.J. Trahan, M. Loveland, N. Davis, E. Ellison, A variational quantum linear solver application to discrete finite-element methods. Entropy, **25**(4), 580 (2023). https://doi.org/10.3390/e25040580
9. Y.Y. Liu et al., Application of a variational hybrid quantum-classical algorithm to heat conduction equation and analysis of time complexity. Phys. Fluids, **34**(11), 117121 (2022). https://doi.org/10.1063/5.0121778
10. Y. Liu et al., *A variational quantum algorithm-based numerical method for solving potential and Stokes flows* (2023). arXiv. https://doi.org/10.48550/ARXIV.2303.01805.
11. H.-L. Liu et al., Variational quantum algorithm for the Poisson equation. Phys. Rev. A, **104**(2), 022418 (2021). https://doi.org/10.1103/PhysRevA.104.022418
12. E. Farhi, J. Goldstone, S. Gutmann, *A quantum approximate optimization algorithm* (2014). arXiv. https://doi.org/10.48550/ARXIV.1411.4028.
13. C. Cîrstoiu, Z. Holmes, J. Iosue, L. Cincio, P.J. Coles, A. Sornborger, Variational fast forwarding for quantum simulation beyond the coherence time. npj Quantum. Inf., 6(1), 82 (2020). https://doi.org/10.1038/s41534-020-00302-0
14. D. Herman et al., Quantum computing for finance. Nat. Rev. Phys., **5**(8), 450–465 (2023). https://doi.org/10.1038/s42254-023-00603-1
15. P. Mocz, A. Szasz, Toward cosmological simulations of dark matter on quantum computers, ApJ, **910**(1), 29 (2021). https://doi.org/10.3847/1538-4357/abe6ac
16. T. Azfar, R. Ke, O.M. Raisuddin, J. Holguin-Veras, *Quantum-assisted vehicle routing: Realizing QAOA-based approach on gate-based quantum computer* (2025). arXiv: arXiv:2505.01614. https://doi.org/10.48550/arXiv.2505.01614
17. M. Lubasch, J. Joo, P. Moinier, M. Kiffner, D. Jaksch, Variational quantum algorithms for nonlinear problems. Phys. Rev. A, **101**(1), 010301 (2020). https://doi.org/10.1103/PhysRevA.101.010301
18. O. Kyriienko, A. E. Paine, V.E. Elfving, Solving nonlinear differential equations with differentiable quantum circuits. Phys. Rev. A, **103**(5), 052416 (2021). https://doi.org/10.1103/PhysRevA.103.052416
19. M. Cerezo, G. Verdon, H.-Y. Huang, L. Cincio, P.J. Coles, Challenges and opportunities in quantum machine learning. Nat. Comput. Sci., **2**(9), 567–576 (2022). https://doi.org/10.1038/s43588-022-00311-3

Part VII

Applications, Future Directions, and Open Problems

This part presents real-world applications, emerging directions, and unresolved challenges in quantum computing, with an emphasis on engineering, scientific computing, and finance. The chapters in this part bridge the theoretical and algorithmic frameworks developed in previous parts to their deployment in practical scenarios.

Chapter 37, "Applications in Engineering and Scientific Computing", surveys proof-of-concept demonstrations and prototype workflows where quantum algorithms have been used for simulation, optimization, and data analysis in engineering contexts.

Chapter 38, "Quantum Machine Learning", introduces the key models, techniques, and practical considerations for applying quantum algorithms to learning and inference tasks, with an emphasis on the interface between quantum and classical computation.

Chapter 39, "Applications in Finance", explores the use of quantum algorithms for problems in financial modeling, including derivatives pricing and portfolio optimization, highlighting both current capabilities and outstanding barriers.

Chapters in this part are intended to provide a realistic appraisal of where quantum computing stands in relation to practical applications, emphasizing not only current achievements but also the open problems and research directions that will define the next phase of progress in the field.

37 Applications in Engineering and Scientific Computing

In the previous part, algorithms for ordinary differential equations and partial differential equations were discussed extensively. Here, we focus on use cases where quantum computers may accelerate numerical solutions, distinguishing between problems that require simulating quantum mechanics and those that do not—roughly separated by nanoscale (quantum) and micro-to-macroscale (classical) problems.

We begin by considering problems involving the simulation of quantum mechanics. Quantum mechanical models provide the most accurate description of many systems, but are typically computationally intractable on classical computers without significant simplifications. A central example is the non-relativistic electronic structure problem, which seeks the minimum eigenvalue (and corresponding eigenstate) of a system of interacting electrons and nuclei, whose Hamiltonian is typically in atomic units:

$$\mathcal{H} = -\sum_{i=1}^{N} \frac{1}{2}\nabla_i^2 - \sum_{A=1}^{M} \frac{1}{2M_A}\nabla_A^2 - \sum_{i=1}^{N}\sum_{A=1}^{M} \frac{Z_A}{r_{iA}} + \sum_{i=1}^{N}\sum_{j>i}^{N} \frac{1}{r_{ij}} + \sum_{A=1}^{M}\sum_{B>A}^{M} \frac{Z_A Z_B}{R_{AB}}$$

where M_A are nuclear masses, Z_A are nuclear charges, N is the number of electrons, and M is the number of nuclei. Using the Born–Oppenheimer approximation to "freeze" the nuclei, one gets the electronic Hamiltonian

$$\mathcal{H}_{elec} = -\sum_{i=1}^{N} \frac{1}{2}\nabla_i^2 - \sum_{i=1}^{N}\sum_{A=1}^{M} \frac{Z_A}{r_{iA}} + \sum_{i=1}^{N}\sum_{j>i}^{N} \frac{1}{r_{ij}}$$

This is a high-dimensional problem, scaling as $3N$. Discretizing the space rapidly becomes infeasible. As an example, for a system with 20 electrons discretized with 10 grid points per spatial dimension, the state space is $\mathcal{O}(10^{60})$.

O. M. Raisuddin and S. De, *Quantum Computing for Engineers*,
https://doi.org/10.1007/978-3-032-03325-3_37

Instead, the electronic structure problem is typically recast in the "second-quantization" framework [1] which is conducive to method development for both classical and quantum computers.

The electronic structure problem enables prediction of key material properties—such as elastic moduli and thermal conductivities—from first principles, and forms the basis of multiscale modeling. Classical approaches like density functional theory (DFT) can treat systems with up to $\sim 10^2 - 10^3$ electrons, but DFT is unreliable for strongly correlated electrons. Thus, quantum electronic structure problems are among the most likely early beneficiaries of quantum advantage, as quantum computers are natively suited to such tasks.

We now turn to micro-macroscale problems, generally governed by classical (often non-unitary) differential equations. Here, quantum algorithms for differential equations can potentially provide a speedup. However, loading classical data into a quantum computer (state preparation) and extracting results (readout) present significant data movement bottlenecks.

For example, in a finite element boundary value problem, one can either (a) discretize classically and upload the data to a quantum computer, or (b) construct the discretized problem directly on the quantum device. The latter is preferable from a data movement and resource standpoint. Nonetheless, Shannon entropy fundamentally limits how much information can be encoded and loaded from classical descriptions.

It is also critical to noting that quantum computers are suited for computing select properties or functionals of the solution, not the entire solution vector. While much work has focused on preparing quantum states representing the solution, efficient extraction of useful observables from these states remains a significant challenge.

One particularly interesting application of quantum computers is in multiscale modeling. In multiscale workflows spanning quantum to macro scales, a quantum computer may accelerate nanoscale computations—e.g., providing material properties to classical solvers at larger scales. Alternatively, one can envision end-to-end quantum workflows, minimizing data movement, though such scenarios are still aspirational.

Reference

1. A. Szabo, N.S. Ostlund, *Modern quantum chemistry: introduction to advanced electronic structure theory* (Dover Publications Inc., Mineola, New York, 2012)

38 Quantum Machine Learning

Machine learning (ML) has evolved to become a cornerstone of modern computational methods. Before delving into quantum machine learning (QML), we provide a brief overview of ML and important concepts linking classical ML to QML.

The goal of ML is to develop techniques for computational tasks without explicitly programming the solution. Instead, ML relies on data to "learn" the task.

ML is broadly categorized into three problems:

- Supervised learning
- Unsupervised learning
- Reinforcement learning.

Supervised learning uses labeled datasets—pairs of inputs and desired outputs—to train a model. Unsupervised learning trains models on unlabeled data to discover patterns, groupings, and correlations. Reinforcement learning trains an agent to develop a policy for coordinating a task, using feedback from the environment and its current state.

ML models are used for a range of computational tasks, including.

- Classification: e.g., classifying tumors as benign or malignant using medical images.
- Regression: e.g., predicting energy consumption using weather data.
- Clustering: e.g., market segmentation to identify key customer groups.
- Generation: e.g., generate metamaterial cell geometries satisfying a target stress–strain response [1].

O. M. Raisuddin and S. De, *Quantum Computing for Engineers*,
https://doi.org/10.1007/978-3-032-03325-3_38

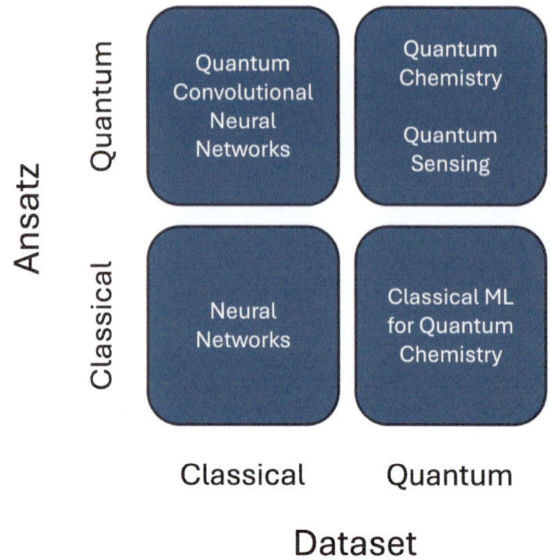

Fig. 38.1 Various combinations of quantum/classical datasets and ansatze with examples

ML models span a broad spectrum, ranging from explainable techniques such as linear regression and support vector machines to more opaque black-box techniques like deep neural networks.

A unifying challenge of ML models is *overparameterization*. When an ML model has too many free parameters, it tends to overfit the data, performing well on training data but poorly on new, unseen data. Overfitting reduces generalizability.

To address overfitting, a combination of regularization and inductive bias is used. Regularization penalizes model complexity, reflecting Occam's razor. Inductive bias refers to the predisposition of certain models toward a subset of possible solutions. For example, convolutional neural networks are well-suited for image processing due to translation invariance, locality, and hierarchical structure.

Major challenges in classical ML include generalizability, overfitting, and underfitting. A growing concern is the computational cost and power consumption, especially for large models—e.g., training modern large language models (LLMs) can consume as much power as a small city.

Quantum machine learning (QML), much like classical machine learning, encompasses a broad array of models and methodologies. However, in QML, there is a foundational distinction that does not appear in the same way in classical ML: Both the model (ansatz) and the data can be either classical or quantum.

- Ansatz here refers to the structure of the model or the functional form of the mapping from input to output (for example, a neural network, a support vector machine, and a parameterized quantum circuit).
- Data refers to the information that the model is trained on or the ground-truth process mapping the inputs to outputs.

This gives rise to several possible scenarios (Fig. 38.1):

1. Classical ansatz, classical data: The standard scenario in classical machine learning (e.g., training a neural network on images).
2. Quantum ansatz, classical data: Quantum models (such as parameterized quantum circuits) trained or used on data originating from classical sources.
3. Classical ansatz, quantum data: Classical models that process data generated from quantum experiments or quantum sensors (less common, but possible).
4. Quantum ansatz, quantum data: Quantum models trained on quantum data—this is considered the most promising avenue for realizing a true quantum advantage, as quantum data can exhibit structures or correlations that are exponentially difficult to represent or process classically.

In summary, the first step in characterizing any QML approach is to specify.

- Whether the data is classical (bitstrings, real-valued vectors, etc.) or quantum (quantum states, results from quantum experiments).
- Whether the model/ansatz is classical (e.g., a traditional neural network) or quantum (e.g., a quantum circuit with trainable gates).

This classification is fundamental because the potential advantages and challenges of QML depend crucially on these choices. For instance, quantum models are expected to provide their greatest benefits when both the data and the model are quantum, due to the exponential complexity of quantum information when represented classically.

We note that the application of various quantum algorithms, such as QLSAs, to machine learning may also be considered instances of QML. In this chapter, we focus on instances of QML that are more complex than a direct application of general quantum algorithms to classical machine learning algorithms. With this motivation, in the remainder of this chapter, we will discuss various aspects and considerations of QML in general, along with various examples and configurations of instances of QML.

We begin our discussion with one of the earliest and most conceptually simple QML instances: quantum support vector machines [2].

Support vector machines are supervised ML models for classifying data by mapping the input data $X : x \in \mathbb{R}^n$ to a higher dimensional space using a kernel $\phi(x) \in \mathbb{R}^m$, where $m \geq n$, and then projecting onto $\mathbb{R}$ by computing the inner product $k : \phi(x_i), \phi(x_j) \in \mathbb{R}$. We denote the labels of each x_i as y_i s.t. $Y : y \in \{+1, -1\}$. This method is referred to as a kernel trick.

A data point can then be classified as

$$y = \mathrm{sgn} \sum_{i=1}^{n} w_i y_i k(x_i, x_j) \in \{+1, -1\}$$

Training a support vector machine with the kernel trick requires computation of the kernel matrix K where $K_{ij} = k(x_i, x_j)$ to optimize w_i.

Since $\phi(\cdot)$ resides in a high-dimensional space, computing the inner products $\langle \phi(x_i), \phi(x_j) \rangle$ may be prohibitively expensive unless the kernel function $\phi(\cdot)$ is

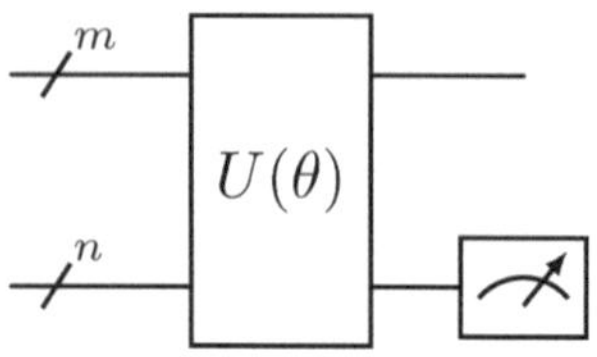

Fig. 38.2 A one-layer quantum neural network

chosen carefully. In a quantum support vector machine, the kernel matrix K is computed using a quantum computer, and w_i are optimized using classical methods. One may compute

$$\langle \phi(x_i), \phi(x_j) \rangle = \langle 0 | U(x_i)^{\dagger} U(x_j) | 0 \rangle$$

where $U(\cdot)$ is a parametrized quantum circuit. This expands the choice of kernels beyond classically tractable kernels.

Next, we consider principal component analysis (PCA), an elementary technique for unsupervised learning. Given a dataset X, where each column of X corresponds to a variable, the columns can be standardized as $\tilde{X}_i = \frac{X_i - \mu_i}{\sigma_i}$, and the covariance matrix can be computed as

$$C = \tilde{X}^T X$$

An eigendecomposition $C = V \Lambda V^T$ yields principal components (the largest eigenvalues and their eigenvectors), reducing the dataset and revealing its main correlations. A quantum algorithm for PCA has been developed [3]. Since the proof uses density matrix formalism, we omit implementation details here.

We now consider the more "general" form of QML: Parametrized quantum circuits with associated cost functions as introduced in Chap. 35: Variational Algorithms: Theory Fig. 38.2.

A key ingredient in QML is encoding data as a quantum state. Beyond basis, amplitude, and phase encoding, other encodings include angle encoding, dense encoding, and feature maps based on Z, ZZ, and Pauli rotations. Data encoding for quantum computations remains an active area of research.

Training parameterized quantum circuits presents several difficulties [4]. Like classical ML models, these circuits can have many local minima, making global optimization NP-hard. Overparameterized circuits may eliminate local minima but risk overfitting.

The optimization landscape of parameterized quantum ansatzes can exhibit barren plateaus (exponentially vanishing gradients) and narrow gorges (exponentially lower global minima) as the parameter count increases. Remedies include using ansatze with greater inductive bias, measuring local rather than global observables, and ensuring measurement qubits are not highly entangled with hidden layers.

Noise in NISQ hardware further complicates optimization; increasing circuit depth amplifies the effects of noise and suppresses meaningful features in the optimization landscape.

Before proceeding further, we state here a remarkable result in quantum machine learning from [5]. Most supervised quantum machine learning models are (equivalent but possibly sub-optimal) quantum kernel models. This profound result can be summarized as follows.

Theorem *Given a data encoding. $|\phi(x)\rangle$ for a classification problem. For most quantum machine learning models, the model obtained using quantum support vector machines is optimal in the m-dimensional subspace spanned by $\phi(\cdot)$ and tractable. Training variational ansatzes on the same embedding $|\phi(x)\rangle$ does not guarantee an optimal model in this subspace, even if the non-convex cost function is globally optimized, which is intractable (NP-hard) in general.*

This result emphasizes the importance of the data embedding $|\phi(x)\rangle$ on the expressibility and performance of QML models. We note that "tractable" here does not imply efficient training or inference.

Recent work has expanded the scope of QML to generative diffusion models. Classical diffusion models are computationally expensive due to an iterative application of a neural network (typically a denoising model like a U-Net). Diffusion models transform a Gaussian noise distribution into a target distribution matching the training dataset. While this process is not necessarily unitary, the manipulation of probability distributions is a particularly interesting and powerful application for quantum computers. Quantum diffusion models aim to replace the denoising model with a trainable quantum circuit [6, 7].

QML may also be a hybrid between classical and quantum models. While QML models are restricted to unitary (linear) transformations and (linear) projections, classical ML models rely heavily on nonlinear operations for expressibility.

QML models typically have the drawback of requiring access to a quantum computer for both training and inference, and quantum computers are a scarce and valuable resource in the NISQ era. To overcome this barrier, a classical "shadow model" approach has been proposed, enabling deployment of trained models on classical computers [8].

QML is an emerging field and is largely driven by experimental and empirical evidence. For in-depth analysis and investigation of the scalability of QML, high-quality hardware is required.

Beyond utilizing quantum computing as a tool for machine learning, machine learning has also been successfully applied to improve the fidelity of quantum computing. Machine learning models have been developed to optimize transpilation and identify more effective error-correction schemes [9].

References

1. J.-H. Bastek, D.M. Kochmann, Inverse design of nonlinear mechanical metamaterials via video denoising diffusion models. Nat. Mach. Intell., **5**(12), 1466–1475 (2023). https://doi.org/10.1038/s42256-023-00762-x

2. D. Anguita, S. Ridella, F. Rivieccio, R. Zunino, Quantum optimization for training support vector machines. Neural Networks., **16**(5–6), 763–770 (2003). https://doi.org/10.1016/S0893-6080(03)00087-X
3. S. Lloyd, M. Mohseni, P. Rebentrost, Quantum principal component analysis. Nature. Phys., **10**(9), 631–633 (2014). https://doi.org/10.1038/nphys3029
4. M. Cerezo, G. Verdon, H.-Y. Huang, L. Cincio, P.J. Coles, Challenges and opportunities in quantum machine learning. Nat. Comput. Sci., **2**(9), 567–576 (2022). https://doi.org/10.1038/s43588-022-00311-3
5. M. Schuld, *Supervised quantum machine learning models are kernel methods (2021)*. arXiv: arXiv:2101.11020. https://doi.org/10.48550/arXiv.2101.11020
6. A. Cacioppo, L. Colantonio, S. Bordoni, S. Giagu, *Quantum diffusion models* (2023). arXiv: arXiv:2311.15444. https://doi.org/10.48550/arXiv.2311.15444
7. M. Kölle, G. Stenzel, J. Stein, S. Zielinski, B. Ommer, C. Linnhoff-Popien, *Quantum denoising diffusion models* (2024). arXiv: arXiv:2401.07049. https://doi.org/10.48550/arXiv.2401.07049
8. S. Jerbi, C. Gyurik, S.C. Marshall, R. Molteni, V. Dunjko, Shadows of quantum machine learning. Nat. Commun., **15**(1), 5676 (2024). https://doi.org/10.1038/s41467-024-49877-8
9. H. Wang et al., *Transformer-QEC: Quantum error correction code decoding with transferable transformers* (2023). arXiv: arXiv:2311.16082. https://doi.org/10.48550/arXiv.2311.16082

39 Applications in Finance

While scientific and engineering computation is notorious for its reliance on large-scale and high-performance computing, the financial industry also contributes significantly to the global computational workload. Quantum computing has the potential to accelerate several problems in finance. In this chapter, we introduce key financial problems where quantum speedups may be possible.

Before presenting computational problems in finance, we introduce foundational concepts in finance. Assets are resources with economic value that can be owned or controlled. These include tangible assets (e.g., gold, wheat, crude oil), and intangible financial instruments such as currency, stocks, options, derivatives, and contracts.

Assets are typically traded on markets, where buyers and sellers determine prices. A financial portfolio is simply a collection of different investments.

Financial derivatives are contracts whose value is “derived” from underlying assets. In its simplest form, the underlying asset for a derivative could be a tangible asset or commodity. A more complex derivative can be a mix of tangible assets, commodities, stocks, options, sub-derivatives, and many other forms of financial instruments. Even a loan contract can be traded, making it a financial instrument in itself. Thus, the term "derivative" broadly covers any combination of financial instruments with monetary value.

All assets involve risk—the possibility that its value may diminish, disappear, or even become a liability. For example, fiat currencies are subject to the stability of the governments backing them, among other factors. An example of an asset with a risk of becoming a liability is a financial contract to receive barrels of crude oil. During the 2020 coronavirus pandemic, the value of oil futures contracts fell below zero due to impending penalties that contract owners would have to pay for disruptions in the oil supply chain leading to an oversupply. Generally, high risk

O. M. Raisuddin and S. De, *Quantum Computing for Engineers*,
https://doi.org/10.1007/978-3-032-03325-3_39

comes with the possibility of high reward. For the same potential return, a rational investor will always prefer lower risk.

To make informed decisions and maximize gains, predicting market trends and optimizing investments are crucial. Better predictions and optimization give market players a competitive advantage. Beyond market prediction and portfolio optimization, financial institutions facilitate a tremendous number of transactions around the clock. To minimize their losses, it is important to detect any fraudulent transactions. These tasks are typically computationally expensive, and quantum computing has the potential to accelerate them. We provide below a brief introduction to these computational problems, and refer readers to a variety of excellent references on these topics [1–7].

Derivatives Pricing and Risk Management

The prices of a derivative's components fluctuate and may be correlated. A stochastic model $S : \{0, 1\}^r \rightarrow \{0, 1\}^n$, e.g., the Black–Scholes model, predicts the state of the market. A payoff function $f : \{0, 1\}^n \rightarrow \mathbb{R}$ evaluates the derivative's value (V) at a given market state:

$$V = E_{x \sim S} f(x)$$

Because these models are high-dimensional and stochastic, analytic solutions are typically infeasible, so Monte Carlo methods are used. Quantum Monte Carlo may be used to model and estimate predicted prices with a theoretical quadratic speedup by encoding S as an algorithm $\mathcal{A}$ and f as a rotation operation $\mathcal{R}$ as discussed in Chap. 25: Quantum Monte Carlo.

This quadratic speedup has several implications. Reducing the number of samples needed to predict the market can potentially allow faster prediction, which is crucial in high-frequency trading. The other implication is that the same number of samples yields a prediction with a quadratically smaller confidence interval, improving the fidelity of the prediction (up to the fidelity of the model). Furthermore, quantum computing may even allow the simulation of more complex and precise models that are simply intractable using classical computers.

Portfolio Optimization

Portfolio optimization seeks the best allocation of assets to maximize return and minimize risk. Let N be the number of assets, with prices $p \in \mathbb{R}^N$, and expected returns $r \in \mathbb{R}^N$. The investments may fluctuate in time and may be correlated with each other, giving rise to the covariance matrix $\Sigma \in \mathbb{R}^{N \times N}$. The asset weights $w \in \mathbb{R}^N$ are chosen subject to.

a budget: $\xi = p^T w$.

expected returns: $\mu = r^T w$.

and risk: $w^T \Sigma w$.

Given a desired return μ with the constraint that all of the budget is invested, a convex quadratic optimization problem can be formed as

$$\min_{w \in \mathbb{R}^N} w^T \Sigma w : \xi = p^T w, \mu = r^T w$$

This can be solved using the method of Lagrange multipliers as the linear system.

$$\begin{bmatrix} 0 \; 0 \; \mu^T \\ 0 \; 0 \; p^T \\ r \; p \; \Sigma \end{bmatrix} \begin{bmatrix} \eta \\ \theta \\ w \end{bmatrix} = \begin{bmatrix} \mu \\ \xi \\ 0 \end{bmatrix}$$

It has been suggested that quantum linear system algorithms may be used to optimize this problem. This variant of portfolio optimization is often referred to as an "unconstrained" optimization problem.

One may impose further constraints, e.g., a positivity constraint to enforce $w_i \geq 0 \; \forall \; i$, i.e., the investor may not sell assets they do not own, which leads to

$$\min_{w \in \mathbb{R}^N} w^T \Sigma w : \xi = p^T w, \mu = r^T w, w_i \geq 0 \; \forall \; i$$

which is a constrained convex optimization problem that can also be solved in polynomial time on a classical computer. However, quantum algorithms have been proposed for a polynomial speedup.

Integer constraints drastically complicate the solution of portfolio optimization problems. As an example, one may impose a constraint on the total number of investments, i.e., the Hamming weight $y = \sum_i \text{bool}(w_i > 0) \leq K$ where bool $x = 0$ for $x = 0$ and bool $x = 1$ otherwise. This leads to an optimization problem

$$\min_{w \in \mathbb{R}^N} w^T \Sigma w : \xi = p^T w, \mu = r^T w, y \leq K$$

which is an NP-hard mixed integer programming problem. One may attempt to approach such combinatorial problems using heuristic methods like QAOA.

Finally, we note that classical machine learning models are often employed for fraud detection, for which quantum machine learning models are an active area of research.

The examples provided in this chapter represent only a small subset of financial problems that may benefit from quantum computing. A wide array of challenges in asset pricing, risk assessment, transaction settlement, option pricing under complex market dynamics, high-frequency trading strategies, and real-time fraud detection could potentially be reformulated for quantum algorithms. However, the fundamental questions remain: Can these formulations be implemented with quantum resources more efficiently than the best classical approaches? And, critically, can quantum computers deliver consistent and actionable value at scale within the constraints and complexities of real-world financial markets? The answer to these questions is not merely academic—it is the billion (or perhaps trillion) dollar question that will ultimately determine the true impact of quantum computing on the financial industry.

References

1. A.S. Naik, E. Yeniaras, G. Hellstern, G. Prasad, S.K.L.P. Vishwakarma, From portfolio optimization to quantum blockchain and security: a systematic review of quantum computing in finance. Financ Innov., **11**(1), 88 (2025). https://doi.org/10.1186/s40854-025-00751-6
2. A. Montanaro, Quantum speedup of Monte Carlo methods. Proc. R. Soc. A., **471**(2181), 20150301 (2015). https://doi.org/10.1098/rspa.2015.0301
3. P. Rebentrost, B. Gupt, T.R. Bromley, Quantum computational finance: Monte Carlo pricing of financial derivatives. Phys. Rev. A, **98**(2), 022321 (2018). https://doi.org/10.1103/PhysRevA.98.022321
4. D. An, N. Linden, J.-P. Liu, A. Montanaro, C. Shao, J. Wang, Quantum-accelerated multilevel Monte Carlo methods for stochastic differential equations in mathematical finance. Quantum, **5**, 481 (2021). https://doi.org/10.22331/q-2021-06-24-481
5. D.J. Egger et al., Quantum computing for finance: state-of-the-art and future prospects. IEEE Trans. Quantum Eng. **1**, 1–24 (2020). https://doi.org/10.1109/TQE.2020.3030314
6. F. Fontanela, A. Jacquier, M. Oumgari, Short communication: A quantum algorithm for linear PDEs arising in finance. SIAM J. Finan. Math., **12**(4), SC98–SC114 (2021). https://doi.org/10.1137/21M1397878
7. R. Orús, S. Mugel, E. Lizaso, Quantum computing for finance: Overview and prospects. Reviews. Phy., **4**, 100028 (2019). https://doi.org/10.1016/j.revip.2019.100028

Uncited References

8. D.P. DiVincenzo, D. Loss, Quantum information is physical. Superlatt. Microstruct., 2(3–4), 419–432 (1998). https://doi.org/10.1006/spmi.1997.0520
9. G.H. Low, N. Wiebe, Hamiltonian simulation in the interaction picture. arXiv:1805.00675. https://doi.org/10.48550/arXiv.1805.00675
10. P.C.S. Costa, D. An, Y.R. Sanders, Y. Su, R. Babbush, D.W. Berry, Optimal scaling quantum linear-systems solver via discrete adiabatic theorem. PRX Quantum, **3**(4), 040303. https://doi.org/10.1103/PRXQuantum.3.040303
11. B. M. Boghosian, W. Taylor, Quantum lattice-gas model for the many-particle Schrödinger equation in d dimensions. Phys. Rev. E, **57**(1), 54–66 (1998). https://doi.org/10.1103/PhysRevE.57.54
12. J. Yepez, Lattice-gas quantum computation. Int. J. Mod. Phys. C, **09**(08), 1587–1596 (1998). https://doi.org/10.1142/S0129183198001436
13. J. Yepez, Quantum lattice-gas model for computational fluid dynamics, Phys. Rev. E, **63**(4), 046702 (2001). https://doi.org/10.1103/PhysRevE.63.046702
14. J. Yepez, Quantum lattice-gas model for the Burgers equation. J. Stat. Phys. **107**(1/2), 203–224 (2002). https://doi.org/10.1023/A:1014514805610
15. G. E. Crooks, Gradients of parameterized quantum gates using the parameter-shift rule and gate decomposition. (2019). arXiv. https://doi.org/10.48550/ARXIV.1905.13311

16. K. Mitarai, M. Negoro, M. Kitagawa, K. Fujii, Quantum circuit learning. Phys. Rev. A, 98(3), 032309 (2018). https://doi.org/10.1103/PhysRevA.98.032309
17. M. Schuld, V. Bergholm, C. Gogolin, J. Izaac, N. Killoran, Evaluating analytic gradients on quantum hardware. Phys. Rev. A, **99**(3), 032331 (2019). https://doi.org/10.1103/PhysRevA.99.032331
18. D. Wierichs, J. Izaac, C. Wang, C. Y.-Y. Lin, General parameter-shift rules for quantum gradients. Quantum, **6**, 677 (2022). https://doi.org/10.22331/q-2022-03-30-677

MIX
Papier aus verantwortungsvollen Quellen
Paper from responsible sources
FSC® C105338
FSC
www.fsc.org

If you have any concerns about our products,
you can contact us on
ProductSafety@springernature.com

In case Publisher is established outside the EU,
the EU authorized representative is:
Springer Nature Customer Service Center GmbH
Europaplatz 3, 69115 Heidelberg, Germany

Printed by Libri Plureos GmbH
in Hamburg, Germany